Solutions Manual
for

Physical Chemistry
A Guided Inquiry

Thermodynamics

James N. Spencer
Franklin & Marshall College

Richard S. Moog
Franklin & Marshall College

John J. Farrell
Franklin & Marshall College

Houghton Mifflin Company Boston New York

Executive Editor: Richard Stratton
Assistant Editor: Danielle Richardson
Editorial Assistant: Rosemary Mack
Manufacturing Manager: Florence Cadran
Senior Marketing Manager: Katherine Greig

Printed in the U.S.A.

ISBN: 0-618-30856-3

23456789 – EB – 07

Contents

Solutions Manual
for

Physical Chemistry
A Guided Inquiry

CA G1 Gases (I)

Exercises

1. a) $PV = nRT \quad P = \frac{nRT}{V}$

   n, T, and V are constant $\therefore$ P is the same in both containers

   b) $ke = \frac{1}{2}\ N m\overline{v^2} \qquad KE = \frac{3}{2n}\ PV = \frac{3}{2n}\ nRT = \frac{3}{2}\ RT$

   T = constant $\therefore$ the average KE is the same in both containers

   c) T = const, the average KE is the same

   $\therefore$ the lighter particles, H_2, must have a greater $\overline{v^2}$ and are, on the average, moving faster

   d) The faster moving particles, H_2, will make more collisions with the walls.

2. $$\sqrt{\left(\frac{J}{K\ mol}\right)\left(\frac{K}{kg/mol}\right)} = \sqrt{\left(\frac{kg\ m^2}{s^2}\right)\left(\frac{1}{kg}\right)} = \sqrt{\frac{m^2}{s^2}} = \frac{m}{s}$$

3. rms speed $= \sqrt{\frac{3RT}{M}}$

   $$rms = \sqrt{\frac{3 \times 8.314 \times 298}{2.016 \times 10^{-3}}} = 1.92 \times 10^3\ \frac{m}{s} \quad \text{for } H_2 \text{ at 298 K}$$

   $$rms = \sqrt{\frac{3 \times 8.314 \times 298}{16.0 \times 10^{-3}}} = 682\ \frac{m}{s} \quad \text{for } CH_4 \text{ at 298 K}$$

4. a) $^{235}UF_6$ (lighter)

   b) $^{235}UF_6$ (makes more collisions w/walls or holes)

5. $d = 1\ \mu m = 1 \times 10^{-6}\ m \qquad \therefore r = 5 \times 10^{-7}\ m \qquad P = 1$ bar

   $V = (4/3)\,\pi\, r^3 = (4/3)\,\pi\,(5 \times 10^{-7}\ m)^3 = 5.2 \times 10^{-19}\ m^3 \quad T = 298$ K

$$n = \frac{PV}{RT} = \frac{(1\text{bar})(5.2\times10^{-19}\ m^3)}{(8.314\ \frac{\text{J}}{\text{K}\cdot\text{mol}})(298K)} = 2.1\times10^{-22}\ \frac{\text{bar}\cdot\text{m}^3}{\text{J}}\ \frac{10^5\ \text{J}}{\text{bar}\bullet\text{m}^3}$$

$n = 2.1\times10^{-17}$ mol

$$2.1\times10^{-17}\ \text{mol}\ \frac{6.022\times10^{23}\ \text{molecules}}{\text{mol}} = 1.3\times10^7\ \text{molecules}$$

6. 1 Pa = 10^{-5} bar = J/m^3

$v_{rms} = 100\ \text{m/s} = \sqrt{\frac{3RT}{M}}$ $\quad$ M = $(6.022\times10^{23})(1\times10^{-25}\ \text{kg}) = 6\times10^{-2}$ kg/mol

$$\therefore (100\ \text{m/sec})^2 = \frac{(3)(8.314\frac{\text{J}}{\text{K}\cdot\text{mol}})\,T}{6\times10^{-2}\ \text{kg/mol}} \Rightarrow T = 24\ \text{K}$$

$n = (10^{23}/6.022\times10^{23}) = 0.166$ mol $\qquad V = 1\times10^{-3}\ \text{m}^3$

$$P = nRT/V = \frac{(0.166)(8.314\frac{\text{J}}{\text{K}\cdot\text{mol}})(24\ \text{K})}{1\times10^{-3}\ \text{m}^3} = 3\times10^4\ \text{Pa}$$

$P = 0.3$ bar

$$\text{KE} = \frac{1}{2}\text{Nm}\ \overline{v}^2 = (1/2)\,(10^{23})\,(10^{-25}\ \text{kg})\,(100\ \text{m/s})^2 = 50\ \text{J}$$

7. $V = 2.00\ \text{L} = 2.00\times10^{-3}\ \text{m}^3 \qquad P = 3.05\ \text{bar} = 3.05\times10^5\ \text{Pa}$

$n = (4.07\times10^{23})/(6.022\times10^{23}\ \text{mol}) = 0.676$ mol

$M = 28.0\times10^{-3}$ kg

$$T = PV/nR = \frac{(3.05\times10^5\ \text{Pa})(2.00\times10^{-3}\ \text{m}^3)}{(0.676\ \text{mol})\left(8.314\ \frac{\text{J}}{\text{K}\bullet\text{mol}}\right)} = 109\ \text{K}$$

$$v_{rms} = \sqrt{\bar{v}^2} = \sqrt{\frac{3RT}{M}} = \left[\frac{(3)\left(8.314\,\frac{\text{J}}{\text{K}\bullet\text{mol}}\right)(109\,K)}{0.0280\ \text{kg/mol}}\right]^{1/2} = 311\ \text{m/s}$$

8. $T = 298$ K

He: $M = 4.00 \times 10^{-3}$ kg

H_2: $M = 2.016 \times 10^{-3}$ kg

$$v_{rms} = \sqrt{\frac{3RT}{M}}$$

He: $v_{rms} = 1.36 \times 10^3$ m/s

H_2: $v_{rms} = 1.92 \times 10^3$ m/s (faster; lighter!)

$\text{KE} = (1/2)\,M\,(v_{rms})^2$

He: KE $= 3.65 \times 10^3$ J

H_2: KE $= 3.65 \times 10^3$ J (same; same temperature!)

9. (a) $v_{rms} = \sqrt{\frac{3RT}{M}}$ $\qquad M = 4.00 \times 10^{-3}$ kg

$$T = 10\ \text{K} \qquad v_{rms} = 250\,\frac{\text{m}}{\text{s}} \times \left(\frac{1\ \text{mile}}{1.6\times10^3\ \text{m}}\right)\left(\frac{3600\ \text{s}}{\text{hr}}\right) = 5.6 \times 10^2\ \text{mile/hr}$$

$$T = 100\ \text{K} \qquad v_{rms} = 790\,\frac{\text{m}}{\text{s}} \times \left(\frac{1\ \text{mile}}{1.6\times10^3\ \text{m}}\right)\left(\frac{3600\ \text{s}}{\text{hr}}\right) = 1.8 \times 10^3\ \text{mile/hr}$$

$$T = 1000\ \text{K} \qquad v_{rms} = 2500\,\frac{\text{m}}{\text{s}} \times \left(\frac{1\ \text{mile}}{1.6\times10^3\ \text{m}}\right)\left(\frac{3600\ \text{s}}{\text{hr}}\right) = 5.6 \times 10^3\ \text{mile/hr}$$

(b) No change; v_{rms} depends on T, not P.

10. $M = 0.0280$ kg $\qquad v_{rms} = \sqrt{\frac{3RT}{M}}$

T(K)	speed of sound (m/s)	v_{rms} (m/s)	$\frac{\text{speed of sound}}{v_{rms}}$
293	344	511	0.673
373	386	576	0.670
773	553	830	0.666
1,273	700	1,060	0.660

speed of sound is approximately (2/3) (v_{rms} of N_2)

11. $F(v) = 4\pi v^2 \left(\frac{m}{2\pi kT}\right)^{3/2} e^{-\frac{mv^2}{2kT}}$

$\frac{dF}{dv} = 0$ at the maximum; this point corresponds to v_{mp}

$$= 4\pi \left(\frac{m}{2\pi kT}\right)^{3/2} \frac{d}{dv}\left[v^2 e^{-\frac{mv^2}{2kT}}\right]$$

$$= 4\pi \left(\frac{m}{2\pi kT}\right)^{3/2} \left[2v\, e^{-\frac{mv^2}{2kT}} + (v^2)\left(\frac{-2mv}{2kT}\right) e^{-\frac{mv^2}{2kT}}\right]$$

$$= 4\pi \left(\frac{m}{2\pi kT}\right)^{3/2} e^{-\frac{mv^2}{2kT}} \left[2v - \frac{mv^3}{kT}\right]$$

This function equals zero when $2v = \frac{mv^3}{kT}$

$v^2 = 2kT/m \qquad v_{mp} = \sqrt{\frac{2kT}{m}}$

$R = kN_A \quad M = m\,N_A$, so $\frac{k}{m} = \frac{R}{M}$

$\therefore\ v_{mp} = \sqrt{\frac{2RT}{M}}$

12. $v_{mp} = \sqrt{\frac{2RT}{M}} = \sqrt{\frac{2 \times 8.314 \times 300}{28.0 \times 10^{-3}}} = 422$ m/s

13. $\bar{v} = \sqrt{\frac{8RT}{\pi M}} = \sqrt{\frac{8 \times 8.314 \times 300}{3.14 \times 28.0 \times 10^{-3}}} = 476$ m/s

14.

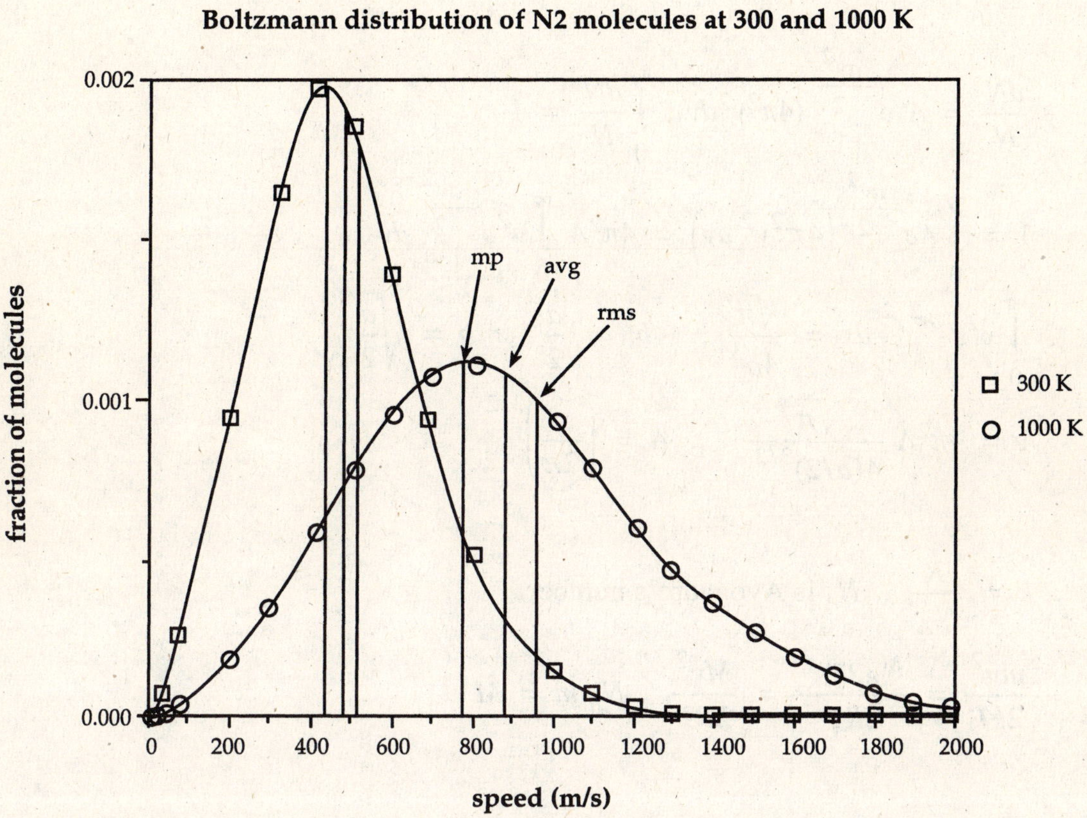

T	v_{rms}	v_{mp}	avg v
300	517	422	476
1000	944	771	870

15. If the curve were "bell-shaped" the average and mp speeds would be equal. But the curve is skewed to the right (to high speeds), so the average speed will be larger than the mp speed.

16. The v_{rms} is greater because the high-speed terms contribute relatively more to Σv^2 than to the Σv.

17. From exercise 13, $v_{avg} = 476\ \frac{m}{s} \times \frac{3600\ s}{hr} \times \frac{100\ cm}{m} \times \frac{in}{2.540\ cm} \times \frac{ft}{12\ in} \times \frac{mi}{5280\ ft}$

$$= 1.06 \times 10^3\ \frac{mi}{hr} \text{ or } \approx 1060\ \frac{mi}{hr}$$

The space shuttle is moving faster.

CA G1A The Maxwell-Boltzmann Speed Distribution Law

1. $$\frac{dN}{N} = A\,e^{-\frac{av^2}{2}}(4\pi^2 v^2 dv);\quad \int_0^\infty \frac{dN}{N} = 1$$

$$1 = \int_0^\infty Ae^{-\frac{av^2}{2}}\left(4\pi^2 v^2\,dv\right) = 4\pi^2 A\int_0^\infty v^2 e^{-\frac{av^2}{2}}\,dv$$

$$\int_0^\infty x^2 e^{-b^2x^2}\,dx = \frac{\sqrt{\pi}}{4b^3}\ ,\quad b^2 = \frac{a}{2}\ ,\quad b = \sqrt{\frac{a}{2}}$$

$$1 = 4\pi^2 A\,\frac{\sqrt{\pi}}{4(a/2)^{3/2}}\ ,\quad A = \left(\frac{a}{2\pi}\right)^{3/2}$$

2. $k = \dfrac{R}{N_A}$, N_A is Avogadro's number

$$\frac{mv^2}{2kT} = \frac{N_A\,mv^2}{2RT} = \frac{Mv^2}{2RT}\ ,\ N_A\,m = M$$

3.

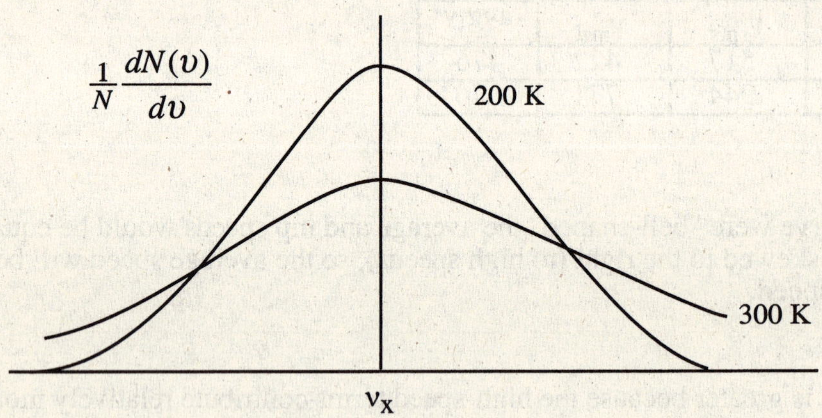

4.

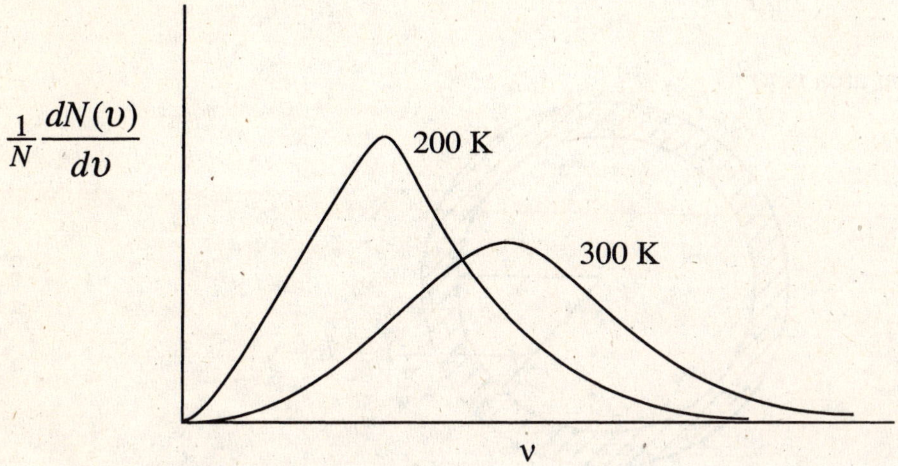

5.

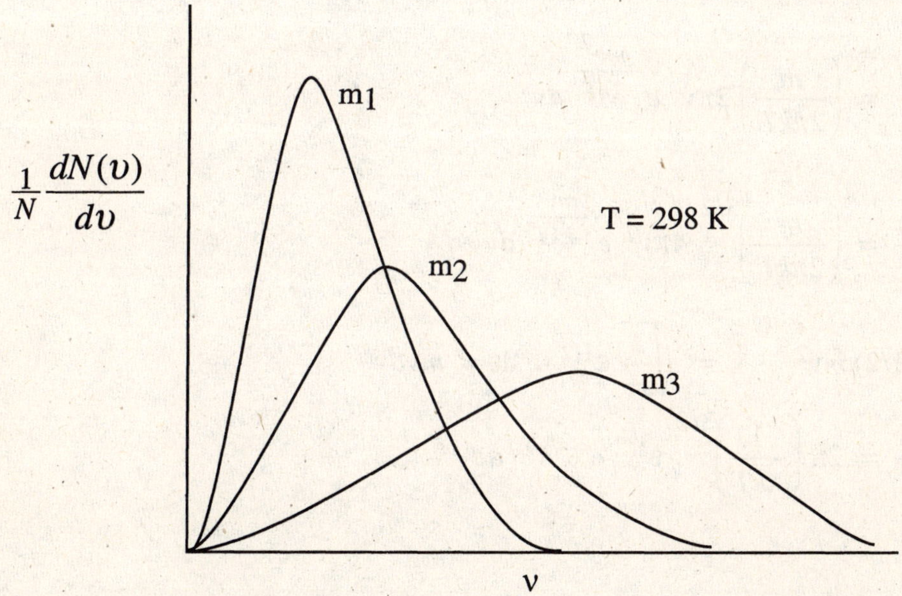

In the above $m_1 > m_2 > m_3$. At a fixed temperature the lighter the mass the more spread out is the velocity distribution.

6. $$\frac{dN(v_x,v_y)}{N} = \left(\frac{m}{2\pi kT}\right) e^{-\frac{m(v_x^2+v_y^2)}{2kT}} dv_x\, dv_y$$

The annular area is πr^2.

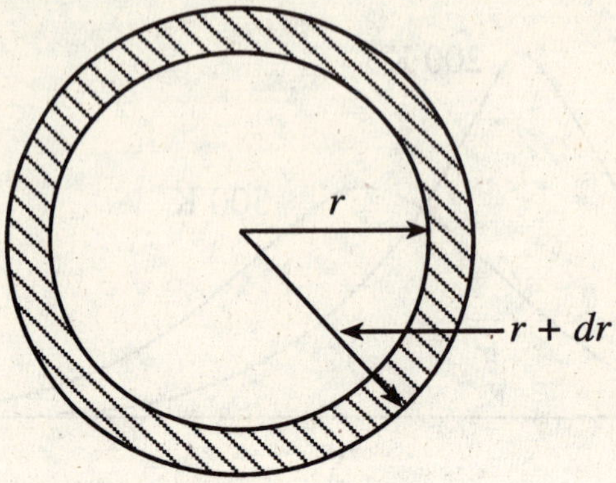

$$\text{d(annular area)} = 2\pi r dr$$

$$dv_x\, dv_y = 2\pi v\, dv$$

$$\frac{dN(v)}{N} = \left(\frac{m}{2\pi kT}\right) 2\pi v\; e^{-\frac{mv^2}{2kT}}\, dv$$

7. $$\frac{dN(v)}{N} = \left(\frac{m}{2\pi kT}\right)^{3/2} 4\pi v^2\, e^{-\frac{mv^2}{2kT}}\, dv$$

$$\varepsilon = (1/2)\, mv^2 \;,\quad v = \sqrt{\frac{2}{m}}\, \varepsilon^{1/2} \;,\quad d\varepsilon = mv dv$$

$$\frac{dN(\varepsilon)}{N} = 2\pi \left(\frac{1}{\pi kT}\right)^{3/2} \varepsilon^{1/2}\, e^{-\varepsilon/\mathrm{kT}}\; d\varepsilon$$

CA G1B Average Values

1. $$\overline{v}^2 = \frac{\int_0^\infty 4\pi N\left(\frac{m}{2\pi kT}\right)^{3/2} v^4 e^{-\frac{mv^2}{2kT}}\,dv}{N}$$

$$= 4\pi\left(\frac{m}{2\pi kT}\right)^{3/2}\int_0^\infty v^4 e^{-\frac{mv^2}{2kT}}\,dv$$

$$\int_0^\infty x^4 e^{-\beta x^2}\,dx = \frac{\sqrt{\pi}}{2}\,\frac{3}{4}\,\frac{1}{\beta^{5/2}}\ ,\quad \beta = m/2kT$$

$$\overline{v}^2 = 4\pi\left(\frac{m}{2\pi kT}\right)^{3/2}\frac{\sqrt{\pi}}{2}\,\frac{3}{4}\,\frac{(2kT)^{5/2}}{m^{5/2}} = 3kT/m\ ,$$

$$\sqrt{\overline{v}^2} = \sqrt{\frac{3kT}{m}}$$

2. $$\left|\overline{v}_x\right| = \frac{N\left(\frac{m}{2\pi kT}\right)^{1/2}\int_0^\infty v_x e^{-\frac{mv_x^2}{2kT}}\,dv_x}{N\left(\frac{m}{2\pi kT}\right)^{1/2}\int_0^\infty e^{-\frac{mv_x^2}{2kT}}\,dv_x}$$

$$\beta = m/2kT\ ,\qquad \int_0^\infty v_x e^{-\frac{mv_x^2}{2kT}}\,dv_x = \frac{kT}{m}\ ,\quad \int_0^\infty e^{-mv_x^2/2kT}\,dv_x = \frac{1}{2}\left(\frac{2\pi kT}{m}\right)^{1/2}$$

$$\left|\overline{v}_x\right| = \frac{2kT/m}{(2\pi kT/m)^{1/2}} = \left(\frac{2kT}{\pi m}\right)^{1/2} = \frac{\overline{v}}{2}$$

3. $$\sqrt{\overline{v}^2} = \sqrt{\frac{3kT}{m}} = \left(\frac{3\times 8.314\times 298}{32.0\times 10^{-3}}\right)^{1/2} = 481\ \text{m/s}\ ,\ \overline{v} = 443\ \text{m/s}\ ,\ v_{mp} = 393\ \text{m/s}$$

4. $$v = \frac{N\left|\overline{v}_x\right|}{2V}$$

It has been shown that (Exercise 2):

$$\left|\overline{v}_x\right| = \frac{\overline{v}}{2}$$

$$v = \frac{N|\bar{v}|}{4V} = \frac{N}{4V}\left(\frac{8kT}{\pi m}\right)^{1/2}$$

$$PV = NkT \ , \ \frac{N}{V} = \frac{P}{kT}$$

$$v = \frac{P}{4kT}\left(\frac{8kT}{\pi m}\right)^{1/2} = \frac{P}{\sqrt{2\pi mkT}}$$

5. $$v = \frac{P}{\sqrt{2\pi mkT}} = \frac{(31.8 \text{ Torr})(133.32 \text{ Pa/Torr})}{\left(\left(2\text{p}\times 1.381\times 10^{-23}\right)(303)\left(\frac{0.01802}{6.022\times 10^{23}}\right)\right)^{1/2}}$$

$$= 1.51 \times 10^{26} \text{ molecules/m}^2\text{s} = 1.51 \times 10^{22} \text{ molecules/cm}^2\text{s}$$

From a surface of 10 cm^2, 1.51×10^{23} molecules/s, which is 4.52 g of water per second.

6. $$v = \frac{P}{\sqrt{2\pi mkT}}$$

$$\frac{v_1}{v_2} = \frac{P_1}{T_1^{1/2}\, m_1^{1/2}} \frac{T_2^{1/2}\, m_2^{1/2}}{P_2}$$

$$\frac{v_{\text{Ne}}}{v_{\text{Ar}}} = \frac{(1/2)}{2^{1/2}}\left(\frac{m_{\text{Ar}}}{m_{\text{Ne}}}\right)^{1/2} = \frac{(1/2)}{2^{1/2}}\left(\frac{40}{20}\right)^{1/2} = 0.50$$

$v_{\text{Ar}} > v_{\text{Ne}}$

CA G2 Gases (II)

1. $PV = nRT$ if T is constant, then $P_1V_1 = P_2V_2$

$P_1 = 0.0465$ bar $\quad P_2 = 11.0$ bar
$V_1 = ?$ $\quad V_2 = 3.00$ L

$$V_1 = \frac{P_2 V_2}{P_1} = \frac{(11.0 \text{ bar})(3.00\ L)}{(0.0465 \text{ bar})} = 7.10 \times 10^2 \text{ L}$$

2. If P is constant, then $(T_1/V_1) = (T_2/V_2)$

$T_1 = 273$ K $\quad T_2 = 373$ K
$V_1 = ?$ $\quad V_2 = 11.4$ L

$$V_1 = \frac{T_1 V_2}{T_2} = \frac{(273 \text{ K})(11.4\ L)}{(373 \text{ K})} = 8.34 \text{ L}$$

3. $n = 0.0440$ moles

P (bar)	*V* (L) 273 K	*V* (L) 373 K
1.000	1.000	1.366
2.000	0.500	0.683
3.000	0.333	0.455
4.000	0.250	0.342
5.000	0.200	0.273
6.000	0.167	0.228
7.000	0.143	0.195
8.000	0.125	0.171
9.000	0.111	0.152
10.000	0.100	0.137

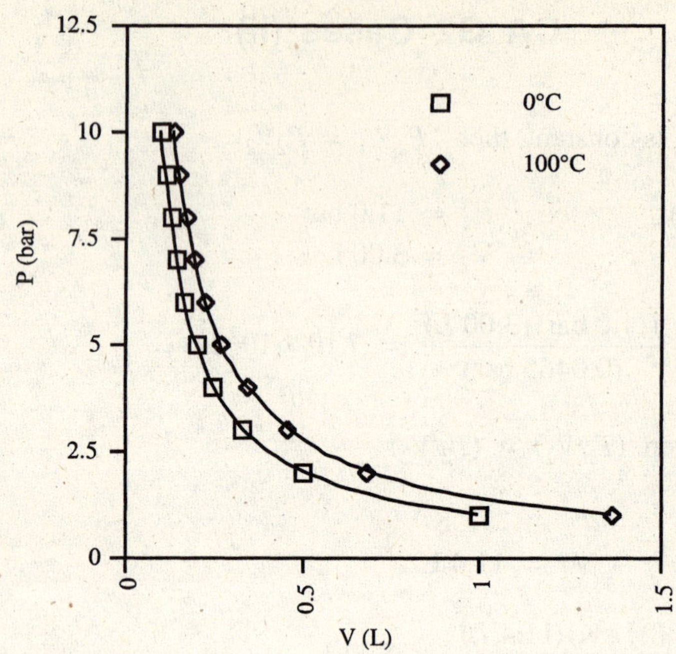

4. $$\frac{P_1 V_1}{T_1} = \frac{P_2 V_2}{T_2}$$

$P_1 = 1.00$ bar $\quad V_1 = 1.00$ L $\quad T_1 = 293$ K

$P_2 = 1.5 \times 10^{-11}$ bar $\quad V_2 = ?$ $\quad T_2 = 1{,}600$ K

$$V_2 = \frac{P_1 V_1 T_2}{P_2 T_1} = \frac{(1.00\text{ bar})(1.00\text{ L})(1{,}600\text{ K})}{(1.5\times10^{-11}\text{ bar})(293\text{ K})} = 3.6 \times 10^{11}\text{ L}$$

5. $PV = nRT \qquad (n/V) = (P/RT)$

mass $= n\,M \;\Rightarrow\; n = mass/M$ (units of M are kg/mole)

$$\therefore\ \frac{n}{V} = \frac{mass}{V \bullet M} = \frac{P}{RT} \qquad d = \frac{mass}{V} = \frac{PM}{RT}$$

6. $M = 0.0280$ kg/mol $\qquad P = 1.00$ bar
$T = 293$ K

$$d = \frac{(1.00\text{ bar})(0.0280\text{ kg/mol})}{(8.314\,\frac{\text{J}}{\text{mol K}})(293\text{ K})} = 1.15 \times 10^{-5}\text{ bar} \bullet \text{kg/J}$$

1 bar $= 10^5$ J/m^3 $\qquad \therefore\ d = 1.15$ kg/m^3

$$d = 1.15\,\frac{\text{kg}}{\text{m}^3}\,\frac{1000\text{ g}}{\text{kg}}\,\frac{1\text{ m}^3}{1000\text{ L}} = 1.15\text{ g/L}$$

7. $P = 0.500$ bar $\qquad d = 0.6565$ g/L $= 0.6565 \times 10^{-5}$ bar kg/J
$T = 293$ K

$$d = \frac{PM}{RT} \Rightarrow M = dRT/P = \frac{(0.6565\times10^{-5}\ \frac{\text{bar kg}}{\text{J}})(8.314\ \frac{\text{J}}{\text{mol K}})(293\ \text{K})}{0.500\ \text{bar}}$$

$M = 0.0320$ kg/mol

8. $$\text{M} = \frac{dRT}{P} = \frac{\left(2.2\times10^{-17}\ \frac{\text{bar kg}}{\text{J}}\right)\left(8.314\ \frac{\text{J}}{\text{mol K}}\right)(1600\ \text{K})}{1.5\times10^{-11}\ \text{bar}} = 0.0195\ \text{kg/mol}$$
$= 19.5$ g/mol

9. The MWs of He, O, and N_2 are 4, 16, and 28 g/mol, respectively. Yes, this mixture seems reasonable because if there were equal amounts of O and N_2 the average MW would be 22.5 g/mol; the presence of He could lower the overall average to 19.5 g/mol. N_2 is more likely to be stable than O_2 because N_2 has a triple bond whereas O_2 has only a double bond.

10. $n_{O_2} = \frac{50.0\ \text{g}}{32.0\ \text{g/mol}} = 1.56$ mol $\qquad n_{CO_2} = \frac{50.0\ \text{g}}{44.0\ \text{g/mol}} = 1.14$ mol

$n_{total} = 1.56 + 1.14 = 2.70$ moles

$$V_{total} = \frac{2.70\ \text{mol} \times 0.08314\ \text{L bar K}^{-1}\ \text{mol}^{-1} \times 293.15\ \text{K}}{0.750\ \text{bar}} = 87.7\ \text{L}$$

$$\frac{P_{O_2}}{P_{total}} = \frac{n_{O_2} \times RT/V}{n_{total} \times RT/V} = \frac{n_{O_2}}{n_{total}} = \frac{1.56\ \text{mol}}{2.70\ \text{mol}} = 0.578$$

$$\frac{P_{CO_2}}{P_{total}} = \frac{n_{CO_2} \times RT/V}{n_{total} \times RT/V} = \frac{n_{CO_2}}{n_{total}} = \frac{1.14\ \text{mol}}{2.70\ \text{mol}} = 0.0.422$$

$$X_{O_2} = \frac{1.56\ \text{mol}}{2.70\ \text{mol}} = 0.578 \qquad X_{CO_2} = \frac{1.14\ \text{mol}}{2.70\ \text{mol}} = 0.422$$

11. a) This gas exhibits ideal gas behavior at low pressures (< 10 bar) at all three temperatures given in the graph. Note that $Z = 1$ at 250 – 300 bar for $T = 100°C$ and

25°C, but this appears to be a coincidence as Z must equal one at some point as the volume factor "wins out" at very high pressures.

b) This gas exhibits ideal gas behavior at higher temperatures: the curve is flatter and Z is closer to 1. The more KE the molecules have, the more they can overcome the intermolecular forces.

c) Low pressure and high temperatures.

12. $Z > 1$ at very high pressures because the volume of the molecules is occupying a significant fraction of the volume of the container; The "b" factor is responsible.

13. $Z < 1$ at moderately high pressures because of the attraction between molecules (which tends to make PV too small. The "a" factor is responsible. The "b" factor cannot be responsible because the volume of the molecules does not occupy a significant fraction of the container at low pressures.

14. (a) $P = \dfrac{1\text{ mol} \times 0.08315\text{ L bar K}^{-1}\text{ mol}^{-1} \times 473.15\text{ K}}{0.935\text{ L}} = 42.1\text{ bar}$

(b) $P = \dfrac{nRT}{V - nb} - \dfrac{an^2}{V^2}$

$$P = \frac{1\text{ mol} \times 0.08315\text{ L bar K}^{-1}\text{ mol}^{-1} \times 473.15\text{ K}}{0.935\text{ L} - 1\text{ mol} \times 0.0374\text{ L mol}^{-1}} - \frac{4.25\text{ bar L}^2\text{ mol}^{-2} \times 1\text{ mol}^2}{(0.935\text{ L})^2}$$

$P = 43.8\text{ bar} - 4.9\text{ bar} = 38.9\text{ bar}$

15. (a) chlorobenzene (dipole-dipole) (b) Cl_2 (London Dispersion forces)
(c) CH_3CH_2OH (hydrogen bonding) (d) Ar (Lon. Dis. forces)
(e) NH_3 (hydrogen bonding) (f) $HN(CH_3)_2$ (hydrogen bonding)

16. Want the largest molecule in each case.
(a) chlorobenzene (b) Cl_2
(c) CH_3OCH_3 (d) Ar
(e) CH_3NH_2 (f) $N(CH_3)_3$

17. For an ideal gas, PV at 300 K = nRT = 24.93 L bar (one mole).

At low P, gas B has PV = 24.95 L bar. This value is larger than predicted by the ideal gas law and is appropriate for Ne, which should have weaker intermolecular forces than NH_3. Also, at 10.0 bar, PV for A is 23.50 L bar (lower than for an ideal gas) and PV for B is 25.1 L bar (higher than for an ideal gas). This makes sense if A is ammonia and B is Ne.

18. $NH_3(g) + HCl(g) \rightleftharpoons NH_4Cl(g)$

In the reaction above, two molecules become one molecule. The pressure should drop because P is proportional to the number of gaseous molecules.

19. Z should be greater than one because the positive argon ions will repel each other thereby increasing the pressure. (This is the opposite of attractive forces.)

Problem

1. (a) $d/P \rightarrow nRT$ as $P \rightarrow 0$; that is the gas becomes perfectly ideal. Because, $d = \frac{PM}{RT}$, we would expect that $\frac{d}{P} \rightarrow \frac{M}{RT}$.

(b) Graph

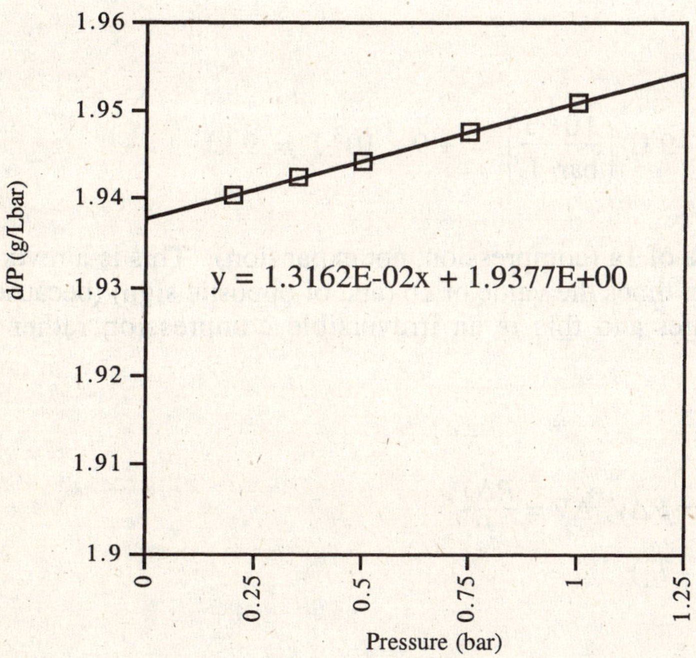

(c) We see that $d/P \rightarrow \frac{1.9377\ \text{g}}{\text{L bar}}$ as $P \rightarrow 0$.

$$\frac{1.9377\ \text{g}}{\text{L bar}} = \frac{44.009\ \text{g/mol}}{R \times 273.15\ \text{K}} \qquad R = 0.083148\ \frac{\text{L bar}}{\text{K mol}}$$

CA T1 Work

Exercises

1. a) $w = -nRT \ln V_f/V_i$

$= -(0.44 \text{ mol})\left(8.314 \frac{\text{J}}{\text{Kmol}}\right)(275\text{K}) \ln\left(\frac{10}{1}\right)$

$= -2.3 \text{ kJ}$

b) $w = -P_{ex}\Delta V = -(1 \text{ bar})(9\text{L}) = -9 \text{ bar}\cdot\text{L}$

1 bar•L = 10^2 J

$= -9 \times 10^2 \text{J}$

2. a) $w = -(0.44 \text{ mol})\left(8.314 \frac{\text{J}}{\text{Kmol}}\right)(275\text{K}) \ln\left(\frac{1}{10}\right)$

$= +2.3 \text{ kJ}$

b) $w = -(10 \text{ bar})(-9 \text{ L})\left(\frac{10^2 \text{ J}}{\text{bar}\cdot \text{L}}\right) = +9 \times 10^3 \text{ J} = 9 \text{ kJ}$

c) 2a is the negative of 1a (compression, not expansion). This is a reversible process. 2b is ten times the value of 1b (and of opposite sign) (because P_{ex} is ten times larger and this is an irreversible compression rather than expansion).

3. a) at constant P,

$dw = -PdV, w = P\Delta V, \Delta V = \frac{R\Delta T}{P}$

$\therefore w = -R(T_2 - T_1)$

b) $PV - \alpha P = RT \Rightarrow P = \frac{RT}{V-\alpha}$

$\therefore dw = -P_{ex}dV = -RT \times \frac{dV}{V-\alpha}$

$w = \int dw = -\int_{V_i}^{V_f} RT \frac{dV}{V-\alpha} = -RT \ln\left(\frac{V_f-\alpha}{V_i-\alpha}\right)$

CA T2 The First Law of Thermodynamics

Exercises

1. $\Delta U_{sys} = q + w = 0 - 200\text{ J} = -200\text{ J}$

2. $\Delta U_{sys} = 0$; U is a state function

3. a) $U_{tot}^{fin} = U_{sys}^{fin} + U_{surr}^{fin}$

   b) $\Delta U_{tot} = \Delta U_{sys} + \Delta U_{surr}$

   c) $\Delta U_{tot} = 0$ because $\Delta U_{sys} = -\Delta U_{surr}$

4. a) $\Delta U_{tot} = 0$

   b) $dU_{tot} = dU_{sys} + dU_{wt} + dU_{surr} = 0$

   $dU_{sys} = -dU_{wt} - dU_{surr}$

   c) $dU_{surr} = -dq$

   d) i) $dU_{sys} = -dU_{wt} - dU_{surr} = dw + dq$

   ii) same as derived in CTQ 8.

5. Some work is possible; if the water warms and expands, work will be done.

6. The temperature rises more in the container with the stops. In the other container some work is done. Assuming the same amount of Ne gas is present in each container, T will rise more in the container with stops because none of the energy transferred to the gas as heat will be expended in raising the weight. No work is done, so no energy is used and the temperature will increase more.

7. $dU = -P_{ex}dV$ $\quad$ $dU < 0$
   $dq = 0$, $dV > 0$ for an expansion

8. $dU = dq$ $\quad$ $dU > 0$
   $dw = 0$, constant volume

CA T3 Enthalpy

Exercises

1. $$C(s) + \frac{1}{2}O_2(g) \longrightarrow CO(g) \qquad \Delta_r H^\circ = -110.53 \text{ kJ/mol}$$
$$H_2O(g) \longrightarrow H_2(g) + \frac{1}{2}O_2(g) \qquad \Delta_r H^\circ = +241.82 \text{ kJ/mol}$$
$$C(s) + H_2O(g) \longrightarrow CO(g) + H_2(g) \qquad \Delta_r H^\circ = +131.29 \text{ kJ/mol}$$

2. $$\Delta_r H^\circ_{298} = 3(-285.830) - (-824.2 \text{ kJ/mol})$$
$$= -33.3 \text{ kJ/mol}$$

3. 0

4. –36.40 kJ/mol, HBr(g) is formed from the elements that compose it, each in their stable state of aggregation.

5. –314.43 kJ/mol, $NH_4Cl(s)$ is formed from its elements as in 4) above. The enthalpy change is the enthalpy of formation.

6. a) $$\Delta_r H = (0.200 \text{ mol}_{N_2} \text{ reacted})(-92.2 \text{ kJ/mol}_{N_2})$$
$$= -18.4 \text{ kJ}$$
$$\text{or } \Delta_r H = (0.200 \text{ mol}_{N_2})\left(-92.2 \frac{\text{kJ}}{\text{mol}_{rxn}} \times \frac{\text{mol}_{rxn}}{1\text{mol}_{N_2}}\right)$$

   b) 0.600 mol H_2

   c) 0.400 mol NH_3

   d) $$\text{For (b) } \Delta_r H = (0.600 \text{ mol } H_2)\left(-92.2 \frac{\text{kJ}}{3 \text{ mol } H_2}\right)$$
$$= -18.4 \text{ kJ}$$
$$\text{For (c) } \Delta_r H = (0.400 \text{ mol } NH_3)\left(-92.2 \frac{\text{kJ}}{2 \text{ mol } NH_3}\right)$$
$$= -18.4 \text{ kJ}$$

   $\Delta_r H$ is the same because of stoichiometry.

CA T3A Enthalpy

Exercises

1. $q = +500 \text{ J}$

   $\Delta H = +500$ J because $\Delta H = q$ for constant P processes.

   $$w = -P_{ex}\Delta V = -(1.00 \text{ bar})(1.43 \text{ L})\left(\frac{10^2 \text{ J}}{\text{bar} \cdot \text{L}}\right) = -143 \text{ J}$$

   $\Delta U = q + w$

   $\Delta U = \Delta H - P\Delta V = 500 \text{ J} - 143 \text{ J} = 357 \text{ J}$

2. a) $\Delta_r n = +1 \qquad \Delta_r(PV) = \Delta_r n \cdot RT = (1)\left(8.314 \frac{\text{J}}{\text{Kmol}}\right)(298 \text{ K})$

   $= 2.48$ kJ/mole of reaction

   $w = -P\Delta V = -2.48 \text{ kJ}$

   $\Delta_r U = \Delta_r H - \Delta_r(PV) = 0 - 2.48 \text{ kJ/mole} = -2.48 \frac{\text{kJ}}{\text{mol}}$

   $\Delta_r H = 0 = q_P$

   b) $\Delta_r n$, $\Delta_r(PV)$, w are all the same as in 2a)

   $\Delta_r U = \Delta_r H - \Delta_r(PV)$

   If $\Delta_r H > 0$, and $\Delta_r(PV) > 0$, then it is not possible to know if $\Delta_r U$ is positive or negative without knowing the magnitudes.

   If $\Delta_r H < 0$, and $\Delta_r(PV) > 0$, then $\Delta_r U < 0$.

   Note that in 2a) $\Delta_r U < 0$ but T is constant. In order for this to be true, heat must flow into the system so that there is no change in U_{tot}.

3. $$\Delta_r(PV) = (1.00 \text{ bar})(2.50 \text{ L})\left(\frac{10^2 \text{ J}}{\text{bar} \cdot \text{L}}\right) = 250 \text{ J} = \frac{0.250 \text{ kJ}}{\text{mole of reaction}} = \Delta_r nRT$$

   $$\Delta_r n \cdot R = (+1 \text{ mole})\left(8.314 \frac{\text{J}}{\text{mol K}}\right) = 8.314 \frac{\text{J}}{\text{K}}$$

   $$T = \frac{\Delta_r(PV)}{\Delta_r n \cdot R} = \frac{250 \text{ J}}{8.314 \text{ J/K}} = 30.1 \text{ K}$$

   $w = -P\Delta V = -0.250 \text{ kJ} = -\Delta_r nRT$

   $q = -1.00 \text{ kJ}$

$$\Delta_r U = \Delta_r H - \Delta_r(PV) = -1.00 \frac{\text{kJ}}{\text{mol}} - 0.250 \frac{\text{kJ}}{\text{mol}} = -1.25 \frac{\text{kJ}}{\text{mol}}$$

4. $\Delta_r n = 0$ adiabatic, $q = 0$, $\Delta_r H = 0$.

   Thus, no change in T; no change in V

   $\Delta V = 0$; $w = 0$

   $\Delta_r U = \Delta_r H - \Delta_r(PV) = 0 - 0 = 0$

5. $H = U + PV \qquad U = H - PV$

   $\Delta_r U = \Delta_r H - \Delta_r(PV)$

   At constant P: $\Delta_r U = \Delta_r H - P\Delta_r V$

6. $\Delta_r U° = -3623$ kJ/mol; $\Delta_r H° = \Delta_r U° + \Delta_r nRT = -3627$ kJ/mol

7. $\Delta_r H°$ represents the difference in total bond strengths when reactants are converted into products under standard conditions. $\Delta_r H° = \Sigma\Delta_r H_f^\circ$ (products) – $\Sigma\Delta_r H_f^\circ$ (reactants).

   $\Delta H°$ is the change in enthalpy when some system undergoes a conversion from an initial state to a final state. $\Delta H° = H°_{final} - H°_{initial}$.

8. $w = -P_{ex}\Delta_r V = -(\Delta_r n)RT$

   so, knowing $\Delta_r n$ and T, w can be calculated.

CA T4 Heat Capacity

Exercises

1. $dU = \left.\frac{\partial U}{\partial T}\right)_V dT + \left.\frac{\partial U}{\partial V}\right)_T dV = \overline{C}_V dT$ since $\left.\frac{\partial U}{\partial V}\right)_T = 0$

$dU = dq + dw = dq - PdV$

$\therefore\ dq = dU + PdV$

$dq = \overline{C}_V dT + PdV = \overline{C}_V dT + RdT$ at constant P

$\frac{dq}{dT} = \overline{C}_P = \overline{C}_V + R \Rightarrow \overline{C}_P - \overline{C}_V = R$

$R = 8.314 \frac{\text{J}}{\text{Kmol}}$

For He, Ne $\overline{C}_P - \overline{C}_V = 20.79 - 12.47 = 8.32 \frac{\text{J}}{\text{Kmol}}$

2. $dq_P = \overline{C}_P dT = \left(50.50 \frac{\text{J}}{\text{Kmol}}\right)(15\text{ K}) = 758 \frac{\text{J}}{\text{mol}}$

3. a)

$$\Delta H = (44.22)(300\text{ K})$$
$$+ \left(\frac{8.79 \times 10^{-3}}{2}\right)(573^2 - 273^2)$$
$$- (-8.62 \times 10^5)\left(\frac{1}{573} - \frac{1}{273}\right)$$
$$= 1.27 \text{ x } 10^4 \text{ J}$$

b)

$$\Delta U = (35.91)(300\text{ K})$$
$$+ \frac{8.79 \times 10^{-3}}{2}(573^2 - 273^2)$$
$$+ 8.62 \times 10^5\left(\frac{1}{573} - \frac{1}{273}\right)$$
$$= 1.02 \times 10^4 \text{ J}$$

4. $\Delta H = a\,\Delta T + \frac{b}{2}\,(T_2^2 - T_1^2) - \left(\frac{c}{T_2} - \frac{c}{T_1}\right)$ but $c = 0$

$T_2 = 313\text{ K} \qquad T_1 = 298\text{ K}$

$$\Delta H = \left(45.94\,\frac{\text{J}}{\text{mol K}}\right)(15\text{ K}) + \frac{16.32 \times 10^{-3}}{2}\,(313^2 - 298^2)$$

$$= 764\text{ J}$$

5. a) $$dH = \left.\frac{\partial H}{\partial T}\right)_P dT + \left.\frac{\partial H}{\partial P}\right)_T dP$$

$$= \overline{C_P}\,dT + \left.\frac{\partial H}{\partial P}\right)_T dP$$

b) $$H = U + PV = U + nRT$$

$$\left.\frac{\partial H}{\partial P}\right)_T = \left.\frac{\partial U}{\partial P}\right)_T + \left.\frac{\partial(nRT)}{\partial P}\right)_T$$

$$= 0 + 0 = 0$$

c) $$\Delta H = \int dH = \int C_P\,dT + 0 = \int C_P\,dT$$

$= C_P\Delta T$ if C_P is not a function of T.

6. $H = U + PV = U + nRT$. Since $U = U(T)$, then $H = H(T)$. Thus, H is a function of T only for an ideal gas.

CA T5 Temperature Dependence of the Enthalpy of Reaction

Exercises

1. $\Delta_r H^\circ_{263} = \Delta_r H^\circ_{273} + \Delta_r \overline{C}_P \cdot \Delta T$

$$= -6025 \frac{\text{J}}{\text{mol}} + \left[36.9 \frac{\text{J}}{\text{Kmol}} - 74.7 \frac{\text{J}}{\text{Kmol}}\right] [-10 \text{ K}]$$

$$= -5.6 \text{ kJ/mol}$$

2. At 298 K: $\Delta_r H^\circ = -\Delta_r H^\circ_f (H_2O(g)) = +241.82$ kJ/mol

$$\Delta_r H^\circ_{500} = \Delta_r H^\circ_{298} + \int_{298}^{500} \Delta_r \overline{C}_P \cdot dT$$

$$\Delta_r \overline{C}_P = \overline{C}_P (H_2(g)) + \frac{1}{2} \overline{C}_P (O_2(g)) - \overline{C}_P (H_2O(g))$$

$$= \left[27.28 + 3.26 \times 10^{-3} \cdot T + \frac{5.0 \times 10^4}{T^2}\right]$$

$$+ \frac{1}{2} \left[29.96 + 4.18 \times 10^{-3} \cdot T - \frac{1.67 \times 10^5}{T^2}\right] - [30.54 + 10.29 \times 10^{-3} T]$$

$$= 11.72 - 4.94 \times 10^{-3} \cdot T - \frac{3.35 \times 10^4}{T^2}$$

$$\Delta_r H^\circ_{500} = 241.82 \frac{\text{kJ}}{\text{mol}} + \int_{298}^{500} \left(11.72 - 4.94 \times 10^{-3} T - \frac{3.35 \times 10^4}{T^2}\right) dT$$

$$= 241.82 \frac{\text{kJ}}{\text{mol}} + 11.72\, T \Big|_{298}^{500} - \frac{1}{2} (4.94 \times 10^{-3}) T^2 \Big|_{298}^{500} + \frac{3.35 \times 10^4}{T} \Big|_{298}^{500}$$

$$= 241{,}820 \frac{\text{J}}{\text{mol}} + 11.72 \frac{\text{J}}{\text{mol} \cdot \text{K}} (202 \text{ K}) - 2.47 \times 10^{-3} (500^2 - 298^2)$$

$$+ 3.35 \times 10^4 \left(\frac{1}{500} - \frac{1}{298}\right) = 243.7 \frac{\text{kJ}}{\text{mol}}$$

3. Assume $\overline{C_P}$ is independent of T over this small T range.

$$\text{363 K:}\quad C_6H_6(g) + \frac{15}{2}O_2(g) \xrightarrow{\Delta_r H^\circ_{363}} 6CO_2(g) + 3H_2O(g)$$

$$\downarrow \left(82\ \frac{J}{mol\ K}\right)(-65\ K) \qquad \downarrow \frac{15}{2}\left(30\ \frac{J}{mol\ K}\right)(-65\ K) \qquad \uparrow 6\left(37\ \frac{J}{mol\ K}\right)(65\ K) \qquad \uparrow 3(34)\ (65\ K)$$

$$\text{298 K:}\quad C_6H_6(g) + \frac{15}{2}O_2(g) \xrightarrow{\Delta_r H^\circ_{298} = -3169\ \frac{kJ}{mol}} 6CO_2(g) + 3H_2O(g)$$

$$\Delta_r H^\circ_{363} = \left[\left(82\ \frac{J}{Kmol}\right)(-65K) + \left(\frac{15}{2}\right)\left(30\ \frac{J}{Kmol}\right)(-65K)\right]$$

$$+ \left[-3169\ \frac{kJ}{mol}\right]$$

$$+ \left[(6)\left(37\ \frac{J}{Kmol}\right)(65K) + (3)\left(34\ \frac{J}{Kmol}\right)(65\ K)\right]$$

$$= \left[-20\ \frac{kJ}{mol}\right] + \left[-3169\ \frac{kJ}{mol}\right] + \left[21\frac{kJ}{mol}\right]$$

$$= -3168\ \text{kJ/mol}$$

4. $H_2O(\ell) \longrightarrow H_2O(g)$ $\qquad T_2 = 298 \qquad T_1 = 373$

$$\Delta_r H^{\circ vap}_{H_2O_{298}} = \Delta_r H^{\circ vap}_{H_2O_{373}} + \Delta_r\overline{C_P}\cdot\Delta T \quad \text{assuming } \overline{C_P} \text{ is independent of } T.$$

$$\Delta T = 298 - 373\ K = -75\ K$$

$$\Delta_r\overline{C_P} = \overline{C_P}\,(H_2O(g)) - \overline{C_P}\,(H_2O(\ell)) = 34\,\frac{J}{mol\ K} - 75\,\frac{J}{K\ mol} = -41\ \text{J/Kmol}$$

$$\Delta_r H^{\circ vap}_{H_2O_{298}} = 40{,}700\,\frac{J}{mol} + \left(-41\,\frac{J}{Kmol}\right)(-75\ K)$$

$$= 43.8\ \text{kJ/mol}$$

CA T6 Entropy

Exercises

1. $dS = \dfrac{dq_{rev}}{T}$

$\int dS = \int \dfrac{dq_{rev}}{T}$ but if T is constant then

$\int dS = \dfrac{1}{T} \int dq_{rev}$

Recall that $dU = dq + dw = dq$ if V is constant ($dw = -PdV$)

Thus: $\int dS = \dfrac{1}{T} \int dU$ at constant T,V

$\Delta S = \dfrac{1}{T} \Delta U \Rightarrow \Delta S_{surr} = \dfrac{1}{T} \Delta U_{surr}$

2. $\Delta_r S^\circ_{273} = \dfrac{\Delta_r H^\circ_{273}}{T}$ for reversible processes ($\Delta S_{tot} = 0$)

$= \dfrac{6025 \text{ J/mol}}{273\text{K}} = 22.1 \text{ J/Kmol}$

3. $\Delta S_{tot} = \Delta S_{sys} + \Delta S_{surr} = \Delta_r S^\circ - \dfrac{\Delta_r H^\circ}{T}$

$= 182.4 \dfrac{\text{J}}{\text{Kmol}} - \dfrac{-2{,}808{,}000 \text{ J/mol}}{298\text{K}}$

$= 9.61 \dfrac{\text{kJ}}{\text{Kmol}}$

4. $\Delta S_{tot} = \Delta_r S^\circ - \dfrac{\Delta_r H^\circ}{T} = -163.34 \dfrac{\text{J}}{\text{Kmol}} - \dfrac{-285{,}830 \text{ J/mol}}{298}$

$= [-163.34 + 959] \dfrac{\text{J}}{\text{Kmol}} = +796 \dfrac{\text{J}}{\text{Kmol}}$; reaction is spontaneous

Sign of $\Delta_r S^\circ$ should be < 0 because liquid is formed from gases and also there is a decrease in the number of moles.

5. $\Delta S_{tot} = \Delta_r S° - \dfrac{\Delta_r H°}{T}$

$$= -3.363 \frac{\text{J}}{\text{Kmol}} - \frac{1895 \text{ J/mol}}{298 \text{ K}} = -9.72 \frac{\text{J}}{\text{Kmol}}$$

$\Delta S_{tot} < 0 \Rightarrow$ not spontaneous

Reverse reaction: $\Delta S_{tot} > 0 \Rightarrow$ spontaneous

6. $dU = dq + dw = dq_{rev} - PdV$ for reversible process

but $dq_{rev} = TdS$

$\therefore dU = TdS - PdV = C_V dT$

$$TdS = C_V dT + PdV \Rightarrow dS = C_V \frac{dT}{T} + \frac{P}{T} dV$$

but $\dfrac{P}{T} = \dfrac{R}{V}$ for one mole of ideal gas

so, $dS = C_V \dfrac{dT}{T} + \text{R} \dfrac{dV}{V}$

$$\Delta S = \int dS = C_V \int_{T_1}^{T_2} \frac{dT}{T} + R \int_{V_1}^{V_2} \frac{dV}{V} \quad \text{if } C_V \neq \text{f}(T)$$

$$= C_V \ln \frac{T_2}{T_1} + R \ln \frac{V_2}{V_1}$$

7. $H = U + PV$

$dH = dU + PdV + VdP$

$= [TdS - PdV] + PdV + VdP$

$dH = TdS + VdP = C_P dT$

$dS = \dfrac{C_P dT}{T} - \dfrac{VdP}{T}$ but $\dfrac{V}{T} = \dfrac{R}{P}$ for one mole of ideal gas

$$\Delta S = \int dS = C_P \int_{T_1}^{T_2} \frac{dT}{T} - R \int_{P_1}^{P_2} \frac{dP}{P} \quad \text{assuming } \overline{C_P} \neq \text{f}(T)$$

$$= C_P \ln \frac{T_2}{T_1} - R \ln \frac{P_2}{P_1}$$

8. Isothermal expansion: $\Delta U = 0$; $\Delta H = 0$ for ideal gas

$$\Delta S = -R \ln \frac{P_2}{P_1} = -\left(8.315 \frac{\text{J}}{\text{Kmol}}\right) \ln\left(\frac{1.0}{10}\right)$$

$$= 19 \frac{\text{J}}{\text{Kmol}}$$

9. $$dS = \frac{C_P dT}{T} - \frac{RdP}{P}$$

$$dS = \left[\frac{28.58}{T} + 3.77 \times 10^{-3} - \frac{0.50 \times 10^5}{T^3}\right] dT - R\frac{dP}{P}$$

$$\Delta S = \int dS = \int_{298}^{1500} 28.58 \frac{dT}{T} + 3.77 \times 10^{-3} \int_{298}^{1500} dT - 0.50 \times 10^5 \int_{298}^{1500} \frac{dT}{T^3}$$

$$-R \int_{1}^{32} \frac{dP}{P}$$

$$= 28.58 \ln \frac{1500}{298} + 3.77 \times 10^{-3}\,(1500 - 298) + \frac{0.50 \times 10^5}{2}\left(\frac{1}{1500^2} - \frac{1}{298^2}\right)$$

$$-R \ln \frac{32}{1} = 21.6 \text{ J/Kmol}$$

10. $$\Delta_r S^\circ = \frac{\Delta_r H^\circ}{T} = \frac{30{,}800 \text{ J/mol}}{353\text{K}} = 87.3 \text{ J/Kmol}$$

CA T7 Entropy Changes as a Function of Temperature

Exercises

1. a)

$$dS = \frac{C_P dT}{T} - \frac{V}{T}dP = \frac{C_P dT}{T} \quad \text{if } dP = 0$$

$$\Delta S = \int dS = C_P \int_{273}^{373} \frac{dT}{T} = C_P \ln \frac{T_2}{T_1}$$

$$\Delta S = \left(75.5 \frac{\text{J}}{\text{Kmol}}\right) \ln \frac{373}{273} = 23.6 \text{ J/Kmol}$$

b) Think of the process as taking place as a series of steps:

i) H_2O (s, 0°C, 1 atm) $\longrightarrow$ H_2O(ℓ, 0°C, 1 atm)

ii) H_2O (ℓ, 0°C, 1 atm) $\longrightarrow$ H_2O(ℓ, 100°C, 1 atm)

iii) H_2O (ℓ, 100°C, 1 atm) $\longrightarrow$ H_2O(g, 100°C, 1 atm)

iv) H_2O (g, 100°C, 1 atm) $\longrightarrow$ H_2O(g, 100°C, 0.50 atm)

i) $\Delta_r S = \dfrac{\Delta_r H_{fus}}{T} = \dfrac{6025 \text{ J/mol}}{273.\text{ K}} = 22.1 \text{ J/Kmol}$

ii) $\Delta S = 23.6$ J/Kmol

iii) $\Delta_r S = \dfrac{\Delta_r H_{vap}}{T} = \dfrac{40{,}700 \text{ J/mol}}{373 \text{ K}} = 109 \text{ J/Kmol}$

iv) $\Delta S = -R \ln \dfrac{P_2}{P_1} = -8.315 \dfrac{\text{J}}{\text{Kmol}} \ln\left(\dfrac{.50}{1}\right) = 5.7 \text{ J/Kmol}$

$\Delta_r S$ (total) $= 22.1 + 23.6 + 109 + 5.7 = 160.5$ J/Kmol

2. a)

$$C_P = C_V + R = \frac{5}{2}R$$

$$\Delta H = C_P \Delta T = \frac{5}{2}R\,(75 \text{ K}) = 1.6 \text{ kJ/mol} = q \text{ (constant } P)$$

$$w = -P_{ex}\Delta V = -R\Delta T = -0.62 \text{ kJ/mol}$$

$$\Delta U = \Delta H - P_{ex}\Delta V = 935 \text{ J/mol}$$

$$\Delta S = C_P \ln \frac{T_2}{T_1} = \frac{5}{2} R \ln\left(\frac{373}{298}\right) = 4.7 \text{ J/Kmol}$$

b) Isothermal, so $\Delta U = 0 = \Delta H$

$$w = -\int P_{ex}dV = -RT\int \frac{dV}{V} = -RT\ln(2) = -2.15\ \frac{\text{kJ}}{\text{mol}}$$

$$q = -w = +2.15\ \text{kJ/mol}$$

$$\Delta S = -R\ln\frac{P_2}{P_1} = -R\ln\left(\frac{1}{2}\right) = 5.76\ \text{J/Kmol}$$

c) $q = 0$

$$\Delta H = C_P\Delta T = \frac{5}{2}R(-65\ \text{K}) = -1.4\ \text{kJ/mol}$$

$$\Delta U = C_V\Delta T = \frac{3}{2}R(-65\ \text{K}) = -811\ \text{J/mol}$$

$$w = \Delta U = -811\ \text{J/mol}$$

$$\Delta S = C_P\ln\frac{T_2}{T_1} = \frac{5}{2}R\ln\left(\frac{308}{373}\right) = -3.98\ \text{J/Kmol}$$

Overall: $\Delta U = 935 + 0 - 811 = 124$ J/mol

$\Delta H = 1.6 + 0 - 1.4 = 0.2$ kJ/mol

$\Delta S = 4.7 + 5.76 - 3.98 = 6.5$ J/Kmol

$q = 1.6 + 2.15 + 0 = 3.7$ kJ/mol

$w = -0.62 - 2.15 - 0.81 = -3.58$ kJ/mol

3. a) Isothermal, so $\Delta U = \Delta H = 0$

$$w = -P_{ex}\Delta V = -(1\ \text{bar})(60.3\ \text{L}) = -60.3\ \text{L•bar} = -6.03\ \text{kJ}$$

$$q = -w = 6.03\ \text{kJ}$$

$$\Delta S = -R\ln\frac{P_2}{P_1} = R\ln\frac{V_2}{V_1} = 8.315\ln\left(\frac{67}{6.7}\right) = 19.1\ \frac{\text{J}}{\text{Kmol}}$$

b) $\Delta U = \Delta H = 0$, isothermal

$dw = -P_{ex}dV = -PdV$, reversible

$$w = -\int_{6.7}^{67}\frac{RT}{V}dV = -3(8.314)(273)\ln\frac{67}{6.7}$$

$$= -1.57\times10^4\ \text{J} = -q,\quad q = 5.23\times10^3\ \text{J/mol}$$

$$\Delta S = \frac{q}{T} = \frac{5.23\times10^3}{273} = 19.1\ \text{J/Kmol}$$

4. $\Delta_r\overline{C}_P = 75.5 - 37.7 = 37.8 \dfrac{\text{J}}{\text{mol K}}$ for melting of ice

$\Delta_r H^\circ_{T_2} = \Delta_r H^\circ_{T_1} + \Delta_r\overline{C}_P \cdot \Delta T$ $\qquad$ $\Delta_r H^\circ_{273} = 6025$ J/mol

$\Delta_r H^\circ_{263} = 5647$ J/mol $\qquad$ $\Delta_r H^\circ_{283} = 6403$ J/mol

$$\Delta S_{surr} = \frac{-\Delta_r H^\circ_T}{T}$$

$$\Delta_r S_{sys} = \Delta_r S^\circ_{273} + \Delta_r\overline{C}_P \ln\left(\frac{T}{273}\right)$$

	+10°C	0°C	–10°C
$\Delta_r H_T$ (J/mol)	6403	6025	5647
ΔS_{surr} (J/Kmol)	–22.6	–22.1	–21.5
ΔS_{sys} (J/Kmol)	+23.5	+22.1	+20.7
ΔS_{tot} (J/Kmol)	+0.9	0	–0.8

5. a) $\Delta T = 0 \Rightarrow \Delta U = \Delta H = 0,\ n = \dfrac{PV}{RT} = 4.41$ mol

$$\Delta S = -R \ln \frac{P_2}{P_1} = -(8.315) \ln\left(\frac{1}{10}\right) = 19.1 \text{ J/Kmol}$$

$$w = -RT \ln \frac{P_2}{P_1} = -(8.315)\,(273) \ln\left(\frac{1}{10}\right) = 5.23 \text{ kJ/mol}$$

$q = -w = -\,5.23$ kJ/mol

For 4.41 mol, $\Delta S = 84$ J/K, $w = 23.1$ kJ $= -q$

b) Adiabatic, $q = 0$, $P_{ex} =$ constant

$$\Delta U = w = n\overline{C}_V (T_2 - T_1) = -P_2 (V_2 - V_1)$$

$$n\overline{C}_V (T_2 - T_1) = -P_2 \left(\frac{nRT_2}{P_2} - \frac{nRT_1}{P_1}\right)$$

$$T_2 = 175 \text{ K}$$

$$\Delta U = n\overline{C}_V (175 - 273) = -5.39 \text{ kJ}$$

$$\Delta H = n\overline{C}_P\,(175 - 273) = -8.98 \text{ kJ}$$

$$\Delta S = n\int \frac{\overline{C}_P dT}{T} - nR\int \frac{dP}{P}$$

$$= n\,5/2\,R \ln\frac{175}{273} - nR\ln\frac{1}{10}$$

$$= 44 \text{ J/K}$$

6. $dS > dq/T$ spontaneous

$dS = dq/T$ reversible

$$\left.\begin{array}{ll} dS > dq/T & \text{spontaneous} \\ dS = dq/T & \text{reversible} \end{array}\right\} dS \geq \frac{dq}{T}$$

7. $d\Delta_r S = \dfrac{\Delta_r\overline{C}_P}{T}\,dT$

$\Delta_r\overline{C}_P > 0$

If $dT > 0$ (T is raised) then first term $\dfrac{\Delta_r\overline{C}_P dT}{T} > 0$ and $d\Delta_r S$ increases.

CA T8 The Third Law of Thermodynamics

Exercises

1. $$\int_0^{63.14} \overline{C_P}(N_2,s)\,\frac{dT}{T} + \frac{\Delta_r H^\circ_{fus}}{63.14\text{ K}} + \int_{63.14}^{77.32} \overline{C_P}(N_2,\ell)\,\frac{dT}{T} + \frac{\Delta H^\circ_{vap}}{77.32\text{ K}}$$

$$+ \int_{77.32}^{298} \overline{C_P}(N_2,g)\,\frac{dT}{T} = S^\circ_{N_2(g),298}$$

Assume that $S^\circ_0 = 0$.

2. Energy altered: gas, weight, surroundings

   Entropy altered: gas, surroundings

3. Some residual randomness, due to intermolecular hydrogen bonds, exists at absolute zero.

4. Not possible. If $\Delta H^\circ_{vap} < 0$, then $\Delta S^\circ_{vap} = \dfrac{\Delta H^\circ_{vap}}{T} < 0$.
   But this would imply a decrease in entropy in going from liquid to gas at a given T, which is not reasonable.

5. a) $$\Delta_r H^\circ = -393.51\,\frac{\text{kJ}}{\text{mol}} - [0+0] = -393.51\text{ kJ/mol}$$

$$\Delta_r S^\circ = 213.74\,\frac{\text{J}}{\text{Kmol}} - \left[205.14 + 5.74\,\frac{\text{J}}{\text{Kmol}}\right] = 2.86\text{ J/Kmol}$$

b) $$\Delta S_{surr} = \frac{-\Delta_r H^\circ}{T} = \frac{393{,}510\text{ J/mol}}{298\text{K}} = 1321\text{ J/Kmol}$$

$$\Delta S_{sys} = \Delta_r S^\circ = 2.86\text{ J/Kmol}$$

$$\Delta S_{tot} = \Delta S_{sys} + \Delta S_{surr} = 1324\text{ J/Kmol}$$

c) $$\Delta_r H^\circ_{400} = \Delta_r H^\circ_{298} + \Delta_r \overline{C_P}\cdot\Delta T \qquad \Delta_r \overline{C_P} = -0.75\,\frac{\text{J}}{\text{Kmol}}$$

$$= -393.51\,\frac{\text{kJ}}{\text{mol}} + \left(-0.0765\,\frac{\text{kJ}}{\text{mol}}\right) = -393.59\text{ kJ/mol}$$

$$\Delta_r S^\circ_{400} = \Delta_r S^\circ_{298} + \Delta_r \overline{C_P} \ln \frac{T_2}{T_1}$$

$$= 2.86 \frac{\text{J}}{\text{Kmol}} + \left(-0.75 \frac{\text{J}}{\text{Kmol}}\right) \ln\left(\frac{400}{298}\right)$$

$$= 2.64 \text{ J/Kmol}$$

$$\Delta S_{surr} = \frac{-\Delta_r H^\circ}{T} = \frac{393{,}590 \text{ J/mol}}{400\text{K}} = 984 \frac{\text{J}}{\text{Kmol}}$$

$$\Delta S_{tot} = 2.64 \frac{\text{J}}{\text{mol K}} + 984 \frac{\text{J}}{\text{mol K}} = 987 \text{ J/Kmol}$$

d) $w = -P_{ex}\Delta V = -\Delta_r n\, RT$ (at constant T)

But $\Delta_r n = 0$ (no change in number of moles of gas)

$\therefore w = 0;\ \Delta_r U = \Delta_r H - P\Delta_r V = \Delta_r H = -393.51 \frac{\text{kJ}}{\text{mol}}$

e) $\Delta_r S^\circ_0 = 0$, assuming $S^\circ_0 = 0$ for all species at $T = 0$.

CA T9 Gibbs Energy and Helmholtz Energy

Exercises

1. a) At 0°C and 1 bar, $G_{H_2O(\ell)} = G_{H_2O(s)}$.

   b) At 100°C and 1 bar, $G_{H_2O(\ell)} = G_{H_2O(g)}$.

   c) At 20°C, $G_{H_2O(\ell)} = G_{H_2O(g)}$ when the gas pressure is 17.36 Torr. $P_{H_2O} =$ 17.36 Torr.

   d) At 1 bar and −1.86°C, $G_{sol} = G_{solid}$

   where G_{sol} refers to a liquid solution of 1.0 molal glucose in water and G_{solid} refers to a solid solution.

   e) At 1 bar and 50°C, $G_{I_2(s)} = G_{I_2(sol)}$

   where $G_{I_2(sol)}$ refers to a solution with a concentration of 0.078 g I_2 per liter.

   f) Assuming the system is at equilibrium, $G_{CaCO_3(s)} = G_{CO_2(g)} + G_{CaO(s)}$.

2. If $\Delta G > 0 \Rightarrow$ process will not occur.
   $\Delta G = 0 \Rightarrow$ at equilibrium.
   $\Delta G < 0 \Rightarrow$ process is spontaneous (does occur).

3. $\Delta A > 0 \Rightarrow$ process will not occur.

   $\Delta A = 0 \Rightarrow$ at equilibrium.

   $\Delta A < 0 \Rightarrow$ process is spontaneous (does occur).

CA T10 Gibbs Energy as a Function of Temperature and Pressure

Exercises

1. a) $\left.\frac{\partial G}{\partial T}\right)_P = -S$

   $S > 0$, so $\left.\frac{\partial G}{\partial T}\right)_P < 0 \Rightarrow$ as $T\uparrow, G\downarrow$

   b) $\left.\frac{\partial G}{\partial P}\right)_T = V \qquad V > 0$

   $\therefore \left.\frac{\partial G}{\partial P}\right)_T > 0 \Rightarrow$ as $P\uparrow, G\uparrow$

2. a) $\mu_{pure} = \mu^\circ$

   b) $\mu_{H2(pure)} = \mu^\circ_{H_2}$

   $\mu_{H2(mix)} = \mu^\circ_{H_2} + RT \underbrace{\ln(0.5)}_{\text{a negative number}}$

   $\mu_{H2(mix)}$ is lower free energy.

   Thus, gas will flow to the side with the mixture.

   c) In this case, everything is the reverse of part b). The pure side has lower free energy, and gas flows to the pure side.

CA T10A Euler's Criterion

Exercises

1. a)

$$G = U + PV - TS$$
$$dG = dU + PdV + VdP - TdS - SdT$$
But $dU = TdS - PdV$, so
$$dG = VdP - SdT$$

b)

$$A = U - TS$$
$$dA = dU - TdS - SdT$$
$$= TdS - PdV - TdS - SdT = -SdT - PdV$$

2. $dG = VdP - SdT$

$x_2dy_2 \quad x_1dy_1$

$$\therefore \left(\frac{\partial V}{\partial T}\right)_P = -\left(\frac{\partial S}{\partial P}\right)_T \Rightarrow \left(\frac{\partial S}{\partial P}\right)_T = \left(\frac{-\partial V}{\partial T}\right)_P$$

$dA = -SdT - PdV$

$x_1dy_1 \quad x_2dy_2$

$$\therefore -\left(\frac{\partial P}{\partial T}\right)_V = -\left(\frac{\partial S}{\partial V}\right)_T$$

$$\Rightarrow \left(\frac{\partial S}{\partial V}\right)_T = \left(\frac{\partial P}{\partial T}\right)_V$$

3. $dU = TdS - PdV$, and from Exercise 2 $\left(\frac{\partial S}{\partial P}\right)_T = \left(\frac{-\partial V}{\partial T}\right)_P$

$$\left(\frac{\partial U}{\partial P}\right)_T = \left(\frac{T\partial S}{\partial P}\right)_T - P\left(\frac{\partial V}{\partial P}\right)_T$$

$$= -T\left(\frac{\partial V}{\partial T}\right)_P - P\left(\frac{\partial V}{\partial P}\right)_T$$

If $PV = RT$

$$\left(\frac{\partial U}{\partial P}\right)_T = -\frac{RT}{P} + \frac{RT}{P} = 0$$

$$\left.\frac{\partial U}{\partial V}\right)_T = \left.\frac{T\partial S}{\partial V}\right)_T - P\left.\frac{\partial V}{\partial V}\right)_T = T\left.\frac{\partial P}{\partial T}\right)_V - P$$

If $PV = RT$

$$\left.\frac{\partial U}{\partial V}\right)_T = \frac{RT}{V} - P = 0$$

For a VdW gas

$$\left.\frac{\partial U}{\partial V}\right)_T = \frac{R}{V-b} - P = \frac{a}{V^2}$$

$$\left.\frac{\partial U}{\partial P}\right)_T = 0 \quad \text{for an ideal gas}$$

$$\left.\frac{\partial U}{\partial V}\right)_T = 0 \quad \text{for an ideal gas}$$

U is a function of T only for an ideal gas.

4. $H \equiv U + PV$

$$\left.\frac{\partial H}{\partial T}\right)_P = \left.\frac{\partial U}{\partial T}\right)_P + P\left.\frac{\partial V}{\partial T}\right)_P = C_P$$

$$C_P - C_V = \left.\frac{\partial U}{\partial T}\right)_P + P\left.\frac{\partial V}{\partial T}\right)_P - \left.\frac{\partial U}{\partial T}\right)_V$$

For an ideal gas $U = U(T)$ only

$$\overline{C_P} - \overline{C_V} = P\left.\frac{\partial V}{\partial T}\right)_P = R$$

CA T11 Equilibrium

Exercises

1. $$K_P = \frac{(P_{NH_3})^2}{(P_{N_2})(P_{H_2})^3}$$

   Graphite is a solid.

2. $\Delta_r G° = -RT \ln K_P$ $\qquad K_P = \dfrac{(P_{H_2O})^2}{(P_{H_2})^2 (P_{O_2})}$

   $2H_2(g) + O_2(g) \rightleftharpoons 2H_2O(g)$ at 298 K, 1 bar

   $\Delta_r G° = \Delta_r H° - T\Delta_r S°$

   $\Delta_r H° = 2\Delta_r H_f°(H_2O(g)) = -484$ kJ/mol

   $\Delta_r S° = 2S°(H_2O(g)) - S°(O_2(g)) - 2S°(H_2(g))$

   $$= 2\left(189 \frac{J}{mol\ K}\right) - \left(205.14 \frac{J}{mol\ K}\right) - 2\left(131 \frac{J}{mol\ K}\right) = -89 \frac{J}{Kmol}$$

   $$\Delta_r G° = -484 \frac{kJ}{mol} - (298\ K)\left(-89 \frac{J}{mol}\right) = -457 \frac{kJ}{mol}$$

   $$\therefore \ln K = -\frac{-457\ kJ/mol}{(8.315\ J/mol)(298)} = 185\ ;\ K = 1.5 \times 10^{80}$$

   Note: Values of $\Delta_r H_f°$, $\Delta_r S°$ can be found in Standard Tables.

3. $\Delta_r G° = -RT \ln K_P$ $\qquad K_P = P_{CO_2}$

   Use data from Standard Tables to calculate $\Delta_r G°$.

   $\Delta_r G° = \Delta_r G_f^{\circ}(CaO) + \Delta_r G_f^{\circ}(CO_2(g)) - \Delta_r G_f^{\circ}(CaCO_3$ (calcite)

   $$= \left(-604.03 \frac{kJ}{mol}\right) + \left(-394.36 \frac{kJ}{mol}\right) - \left(-1128.79 \frac{kJ}{mol}\right)$$

   $= 130.4$ kJ/mol

   $$\ln K_P = -\frac{\Delta_r G°}{RT} = -\frac{130{,}400\ J/mol}{(8.315\ Kmol)(298\ K)} = -52.6 \Rightarrow K_P = 1.4 \times 10^{-23}$$

4. $P_i = \frac{n_i RT}{V} = C_i RT$ where $C_i = \frac{n_i}{V}\left(\frac{\text{moles}}{\text{L}}\right)$

$\Sigma P_i = P_{tot} = \Sigma n_i \frac{RT}{V} = \frac{RT}{V}\Sigma n_i = \frac{RT}{V} n_{tot}$

$X_i = \frac{n_i}{n_{tot}} \quad n_i = X_i\, n_{tot}$

$P_i = \frac{n_i RT}{V} = \frac{X_i n_{tot} RT}{V} = X_i \frac{n_{tot} RT}{V} = X_i P_{tot}$

$$K_P = \frac{P_C^c}{P_A^a P_B^b} = \frac{(C_C RT)^c}{(C_A RT)^a (C_B RT)^b} = \frac{(C_C)^c}{(C_A)^a (C_B)^b} \times \frac{(RT)^c}{(RT)^a (RT)^b}$$

$$= K_C \times (RT)^{c-(a+b)} = K_C (RT)^{\Delta n}$$

where $K_C = \frac{(C_C)^c}{(C_A)^a (C_B)^b}$ and $\Delta n = c - (a+b)$

$$K_P = \frac{P_C^c}{P_A^a P_B^b} = \frac{(X_C P_{tot})^c}{(X_A P_{tot})^a (X_B P_{tot})^b} = \frac{(X_C)^c}{(X_A)^a (X_B)^b} \times (P_{tot})^{c-(a+b)}$$

$$= K_X \times (P_{tot})^{\Delta n}$$

5. $P_{tot} = 1.00$ bar $\qquad \Delta_r G^\circ = -RT \ln K_P$

$$= -\left(8.315 \frac{\text{J}}{\text{mol K}}\right)(667.95\text{ K}) \ln(.0444)$$

$$= 17.3 \text{ kJ/mole}$$

Assume originally only $COCl_2(g)$ present.

At equil: $P_{CO} = P_{Cl_2} = P \qquad P_{COCl_2} = 1.00 - 2P$

$$K_P = \frac{P_{CO} P_{Cl_2}}{P_{COCl_2}} = \frac{P^2}{1-2P} = 0.0444$$

$$P^2 = .0444 - .0888\,P \Rightarrow P^2 + .0888\,P - .0444 = 0$$

$$P = \frac{-.0888 \pm \sqrt{(.0888)^2 - 4(-.0444)}}{2}$$

$$= \frac{-.0888 \pm .4307}{2} = 0.171 \text{ bar}$$

$P_{CO} = P_{Cl_2} = 0.171$ bar $\qquad P_{COCl_2} = 0.658$ bar

6. $K_P = \dfrac{P_{NO_2}^2}{P_{N_2O_4}}$ $\qquad P_{N_2O_4} = 1.00 \text{ bar} - P_{NO_2}$

If N_2O_4 is 19% dissociated, then for every 100 molecules of N_2O_4 originally present, only 81 remain at equilibrium. Thus $2 \times 19 = 38$ molecules of NO_2 are produced.

$$\therefore \frac{P_{NO_2}}{P_{N_2O_4}} = \frac{38}{81} \quad \text{assuming ideal behavior}$$

$$\text{or } P_{NO_2} = \frac{38}{81} P_{N_2O_4} = 0.469\, P_{N_2O_4}$$

$$\text{so } P_{N_2O_4} = 1.00 \text{ bar} - (0.469\, P_{N_2O_4})$$

$$1.469\, P_{N_2O_4} = 1.00 \text{ bar} \Rightarrow P_{N_2O_4} = 0.681 \text{ bar}$$

$$\therefore P_{NO_2} = 0.319 \text{ bar}$$

$$K_P = \frac{(0.319 \text{ bar})^2}{0.687 \text{ bar}} = 0.149 \text{ bar}$$

7. $C(\text{graphite}) + H_2O(g) \rightleftharpoons CO(g) + H_2(g)$

a) $$\Delta_r H^\circ_{298} = \Delta_r H^\circ_{298} + \Delta_r \overline{C}_P \Delta T \qquad \Delta T = 100 \text{ K}$$

$$\Delta_r H^\circ_{298} = 131 \text{ kJ/mol}$$

$$\Delta_r \overline{C}_P = (29.14) + (28.84) - (8.64) - (33.58) = 15.76 \text{ J/Kmol}$$

$$\Delta_r H^\circ_{398} = 131 \frac{\text{kJ}}{\text{mol}} + \left(15.76 \frac{\text{J}}{\text{Kmol}}\right) (100 \text{ K}) = 133 \text{ kJ/mol}$$

b) $$\Delta_r S^\circ_{398} = \Delta_r S^\circ_{298} + \Delta_r \overline{C}_P \ln \frac{398}{298}$$

$$\Delta_r S^\circ_{298} = 134 \frac{\text{J}}{\text{Kmol}}$$

$$\Delta_r S^\circ_{398} = 134 \frac{\text{J}}{\text{Kmol}} + \left(15.76 \frac{\text{J}}{\text{Kmol}}\right) \ln \frac{398}{298} = 139 \frac{\text{J}}{\text{Kmol}}$$

c) $$\Delta S_{tot} = \Delta_r S^\circ_{398} - \frac{\Delta_r H^\circ_{398}}{T}$$

$$= 139 \frac{\text{J}}{\text{Kmol}} - \frac{133{,}000 \text{ J/mol}}{398 \text{ K}} = -195 \frac{\text{J}}{\text{Kmol}}$$

d) No. $\Delta S_{tot} < 0$ so reaction does not proceed to the right.

e) $\Delta_r U^\circ = \Delta_r H^\circ - \Delta_r(PV) = \Delta_r H^\circ - \Delta_r(nRT)$

assuming ideal behavior for gases and ignoring the molar volume of solid graphite.

Then $\Delta_r(nRT) = \Delta_r n(RT) = (+1)\,RT$

where $\Delta_r n$ refers to change in gaseous moles.

$$\Delta_r U^\circ = 133\ \frac{\text{kJ}}{\text{mol}} - \left(8.315\ \frac{\text{J}}{\text{Kmol}}\right)(398\ \text{K}) = 130\ \text{kJ/mol}$$

f) $\Delta_r G^\circ = \Delta_r H^\circ - T\Delta_r S^\circ$

$$= 133\ \frac{\text{kJ}}{\text{mol}} - (398\ \text{K})\left(139\ \frac{\text{J}}{\text{Kmol}}\right) = 77\ \text{kJ/mol}$$

$\Delta_r G^\circ > 0$ so reaction does not proceed to the right.

g) $$\ln K_P = -\frac{\Delta_r G^\circ}{RT} = \frac{77{,}000\ \text{J/mol}}{\left(8.315\ \frac{\text{J}}{\text{Kmol}}\right)(398\ \text{K})} = -23.3$$

$\therefore\ K_P = 7.6 \times 10^{-11}$

8. $\Delta\overline{G} = \int_{P^\circ}^{P} \overline{V}dP \cong \overline{V}(P - P_o)$ if V does not change with P, which is a reasonable assumption for a solid or liquid.

$= \overline{G} - \overline{G}^\circ = \mu - \mu^\circ$

$V(P-P^\circ)$ is very small

so $\overline{G} \approx \overline{G}^\circ$ and $\overline{G}$ may be replaced by $\overline{G}^\circ$.

$$\Delta_r G = \mu_{CaO} + \mu_{CO_2} - \mu_{CaCO_3} \approx \mu^\circ_{CaO} + \mu^\circ_{CO_2} + RT\ln P_{CO_2} - \mu^\circ_{CaCO_3}$$

$$\Delta_r G = (\mu^\circ_{CaO} + \mu^\circ_{CO_2} - \mu^\circ_{CaCO_3}) + RT\ln P_{CO_2}$$

$$= \Delta_r G^\circ + RT\ln P_{CO_2}$$

9. a) $$K_P = \frac{(P_{NOCl})^2}{(P_{NO})^2(P_{Cl_2})} = \frac{(1.2\ \text{atm})^2}{(5.0 \times 10^{-2}\ \text{atm})^2(3.0 \times 10^{-1}\ \text{atm})} = 1.9 \times 10^3$$

b) Yes. Some NO and Cl would form.

c) Yes. $Q = 1 \quad Q < K_P$. In order to reach equilibrium, Q must increase. Thus more products will form; reaction proceeds to the right.

d) $$Q = \frac{(1.0\ \text{atm})^2}{(5.0 \times 10^{-3}\ \text{atm})^2\ (3.0 \times 10^{-1}\ \text{atm})} = 1.3 \times 10^5$$

$Q > K_P$. Thus, more reactants must form to reach equilibrium. Reaction proceeds to the left.

CA T12 Temperature Dependence of the Equilibrium Constant

Exercises

1. $$\ln \frac{K_2}{K_1} = -\frac{\Delta_r H}{R}\left(\frac{1}{T_2} - \frac{1}{T_1}\right)$$

$K_1 = 870$ $T_1 = 298$ K $\Delta_r H^\circ = -10.38$ kJ/mol

$K_2 = ?$ $T_2 = 328$ K

$$\ln K_2 = +\frac{10{,}380 \text{ J/mol}}{8.315 \dfrac{\text{J}}{\text{Kmol}}}\left(\frac{1}{328} - \frac{1}{298}\right) + \ln(870)$$

$$= -0.383 + 6.768 = 6.385$$

$\therefore K_2 = 593$ at 328 K

2. $$\ln \frac{14.5 \times 10^3}{4.58 \times 10^3} = -\frac{\Delta_r H^\circ}{R}\left(\frac{1}{311} - \frac{1}{295}\right)$$

$$1.15 = -\frac{\Delta_r H^\circ}{8.315 \text{ J/Kmole}}(-1.744 \times 10^{-4})$$

$\Delta_r H^\circ = +54.9$ kJ/mol

$$\Delta_r G^\circ = -RT \ln K = -\left(8.315 \frac{\text{J}}{\text{Kmol}}\right)(295 \text{ K}) \ln (4.58 \times 10^3)$$
$$= -20.6 \text{ kJ/mol}$$

$$\Delta_r G^\circ = \Delta_r H^\circ - T\Delta_r S^\circ \Rightarrow \Delta_r S^\circ = -\left(\frac{\Delta_r G^\circ - \Delta_r H^\circ}{T}\right)$$

$$\Delta_r S^\circ = -\left(\frac{-20.6 \text{ kJ/mol} - 54.8 \text{ kJ/mol}}{295 \text{ K}}\right) = +255 \frac{\text{J}}{\text{Kmol}}$$

3. $$d\Delta_r H^\circ = \Delta_r \bar{C}_P dT$$

$$\Delta_r H^\circ_T - \Delta_r H^\circ_{298} = \Delta_r \bar{C}_P (T - 298)$$

$$\frac{d\ln K}{dT} = \frac{\Delta_r H^\circ}{RT^2}$$

$$d\ln K = \frac{\Delta_r H^\circ}{RT^2}\, dT$$

$$\int_{K_{298}}^{K_2} d\ln K = \int_{298}^{T_2} \left\{ \frac{\Delta_r H^\circ_{298}}{RT^2} + \frac{\Delta_r \bar{C}_P}{RT^2}(T-298) \right\} dT$$

$$\ln \frac{K_2}{K_{298}} = -\frac{\Delta_r H^\circ_{298}}{R}\left(\frac{1}{T_2} - \frac{1}{298}\right) + \frac{\Delta_r \bar{C}_P}{R} \ln \frac{T_2}{298} + \frac{\Delta_r \bar{C}_P}{R}(298)\left(\frac{1}{T_2} - \frac{1}{298}\right)$$

4. a) $$\Delta_r H^\circ_{298} = \left(-241.826 \frac{\text{kJ}}{\text{mol}} + -74.847 \frac{\text{kJ}}{\text{mol}}\right) - \left(-110.523 \frac{\text{kJ}}{\text{mol}} + 3(0)\right)$$

$$= -206.150 \frac{\text{kJ}}{\text{mol}}$$

$$\Delta_r S^\circ_{298} = \left(186.19 \frac{\text{J}}{\text{Kmol}} + 188.72 \frac{\text{J}}{\text{Kmol}}\right) - \left(197.91 \frac{\text{J}}{\text{Kmol}} + 3\,(130.58) \frac{\text{J}}{\text{Kmol}}\right)$$

$$= -214.74 \frac{\text{J}}{\text{Kmol}}$$

$$\Delta_r G^\circ_{298} = \Delta_r H^\circ_{298} - T\Delta_r S^\circ_{298} = -206.150 \frac{\text{kJ}}{\text{mol}} - (298)\left(-214.74 \frac{\text{J}}{\text{mol}}\right)$$

$$= -142.16 \frac{\text{kJ}}{\text{mol}}$$

$$\ln K_P = \frac{-\Delta_r G^\circ}{RT} = \frac{142{,}160 \text{ J/mol}}{\left(8.315 \frac{\text{J}}{\text{Kmol}}\right)(298 \text{ K})} = 57.372$$

$\therefore\ K_P = 8.25 \times 10^{24}$ at 298 K

b) To calculate K_{P500}, recall that:

$$\frac{d\ln K}{dT} = \frac{\Delta_r H^\circ}{RT^2} \Rightarrow \int_{K_{298}}^{K_{500}} d\ln K = \int_{298}^{500} \frac{\Delta_r H^\circ}{RT^2}\, dT$$

$$\text{so } \ln K_{P500} - \ln K_{P298} = \frac{1}{R}\int_{298}^{500} \frac{\Delta_r H^\circ}{T^2}\, dT$$

In this case we must find $\Delta_r H^\circ$ as a function of T and then integrate!

$$\Delta_r H^\circ_T = \Delta_r H^\circ_{298} + \int_{298}^{T} \Delta_r C_P dT$$

$$\Delta_r C_P = \left[\overline{C_P}(CH_4) + \overline{C_P}(H_2O)\right] - \left[\overline{C_P}(CO) + 3\overline{C_P}(H_2)\right]$$

$$\overline{C_P}(CH_4) = 23.64 + 47.86 \times 10^{-3}\, T - 1.92 \times 10^5/T^2$$

$$\overline{C_P}(H_2O) = 30.54 + 10.29 \times 10^{-3}\, T$$

$$\overline{C_P}(CO) = 28.41 + 4.10 \times 10^{-3}\, T - 0.46 \times 10^5/T^2$$

$$3 \times \overline{C_P}(H_2) = 81.84 + 9.78 \times 10^{-3}\, T + 1.50 \times 10^5/T^2$$

$$\Delta_r C_P = -56.07 + 44.27 \times 10^{-3}\, T - 0.88 \times 10^5/T^2$$

$$\Delta_r H^\circ_T = -206.150\,\frac{\text{kJ}}{\text{mol}} + \int_{298}^{T} \left(-56.07 + 44.27 \times 10^{-3}\, T - \frac{0.88 \times 10^5}{T^2}\right) dT$$

$$= -206{,}150 +$$

$$\left(-56.07 \times T \Big|_{298}^{T} + \frac{1}{2} 44.27 \times 10^{-3}\, T^2 \Big|_{298}^{T} + \frac{0.88 \times 10^5}{T} \Big|_{298}^{T}\right)$$

$$= -206{,}150 - 56.07 \times T + 16{,}709 + 0.02214\, T^2 - 1{,}966 + \frac{0.88 \times 10^5}{T} - 295$$

$$= -191.7\,\frac{\text{kJ}}{\text{mol}} - 56.07\, T + 0.02214\, T^2 + \frac{0.88 \times 10^5}{T}$$

$$\ln K_{P500} = \ln K_{P298} + \frac{1}{R} \int_{298}^{500} \frac{\Delta_r H^\circ}{T^2}\, dT$$

$$= 57.372 + \frac{1}{R} \int \left(-\frac{191{,}700}{T^2} - \frac{56.07}{T} + .02214 + \frac{0.88 \times 10^5}{T^3}\right) dT$$

$$\ln K_{P500} = 57.372$$

$$+ \frac{1}{R}\left[\left(\frac{191{,}700}{T}\Big|_{298}^{500} - 56.07 \ln T\Big|_{298}^{500} + 0.02214\, T\Big|_{298}^{500} - \frac{1}{2}\frac{0.88 \times 10^5}{T^2}\Big|_{298}^{500}\right)\right]$$

$$= 57.372 + \frac{1}{8.315}\ [(-226.3) - (29.0) + (4.5) - (-0.3)]$$

$$= 57.372 + \frac{1}{8.315}\ [-250.5] = 27.246$$

$$\therefore K_{P_{500}} = 6.80 \times 10^{11}$$

5. $$\left.\frac{\partial(\Delta A/T)}{\partial T}\right)_V = \frac{T\left.\frac{\partial \Delta A}{\partial T}\right)_V - \Delta A}{T^2} = \frac{-T\Delta S - \Delta A}{T^2}$$

$$= \frac{-\Delta U}{T^2}$$

$$A \equiv U - TS$$

$$dA = -PdV - SdT$$

$$\left.\frac{\partial A}{\partial T}\right)_V = -S$$

CA T13 Temperature and Pressure Dependence of Phase Equilibria for Pure Phases

Exercises

1. $\left.\frac{\partial G}{\partial T}\right)_P = -S$. Thus, for a given P, the value of G is increased when T is decreased.

   Because S for liquids is less than S for gases, this effect is greater for gases.

   Consider Figure 1, but at a lower T than 373 K.

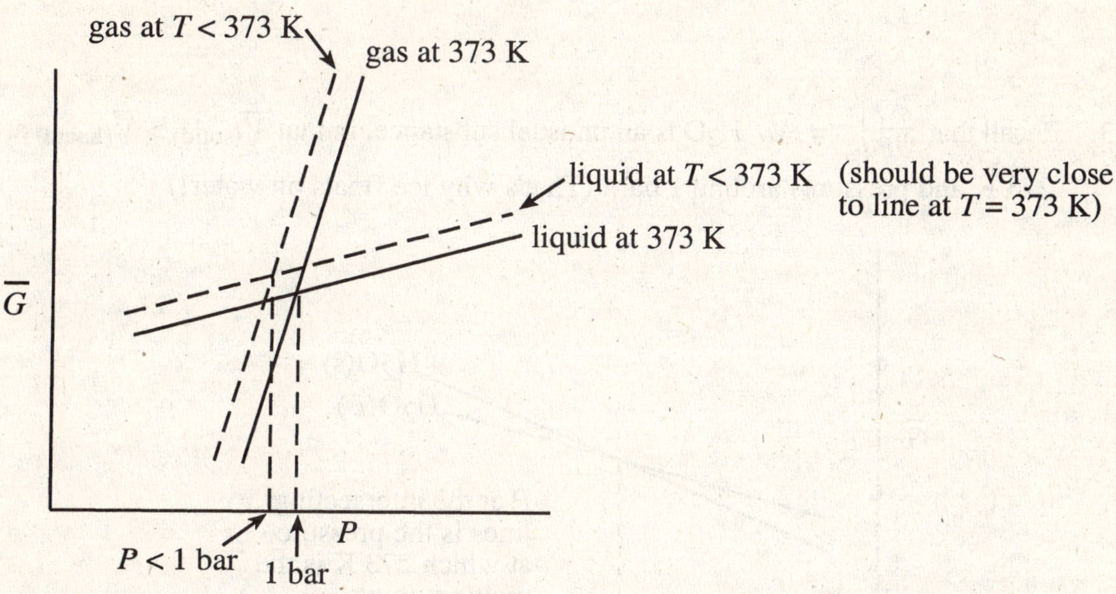

   The intersection of the two dotted lines (T<373) will be at a lower P than 1 bar. Thus, the boiling point at pressures less than 1 bar correspond to T<373 K.

2. $G = H - TS$, if the gas is ideal, H is not a function of pressure. At a fixed temperature, G will vary with P as S varies with P. S changes with P in a logarithmic fashion.

   The line in Figure 1 has a slope of $\left.\frac{\partial G}{\partial P}\right)_T = \overline{V}$. But, for an ideal gas, $\overline{V}$ is not constant as pressure changes: $\overline{V} = \frac{RT}{P}$. As $P\uparrow$, $\overline{V}\downarrow$, so slope decreases.

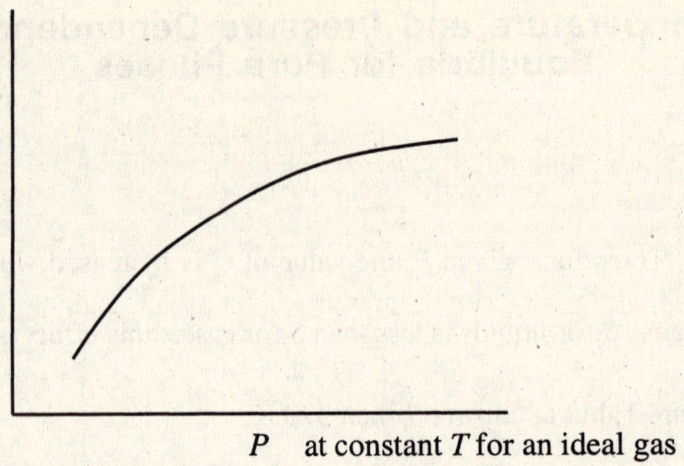

3. Recall that $\left(\frac{\partial \overline{G}}{\partial P}\right)_P = \overline{V}$. H_2O is an unusual substance, in that $\overline{V}_{(solid)} > \overline{V}_{(liquid)}$ at 273 K and pressures around 1 bar. (That's why ice floats on water!)

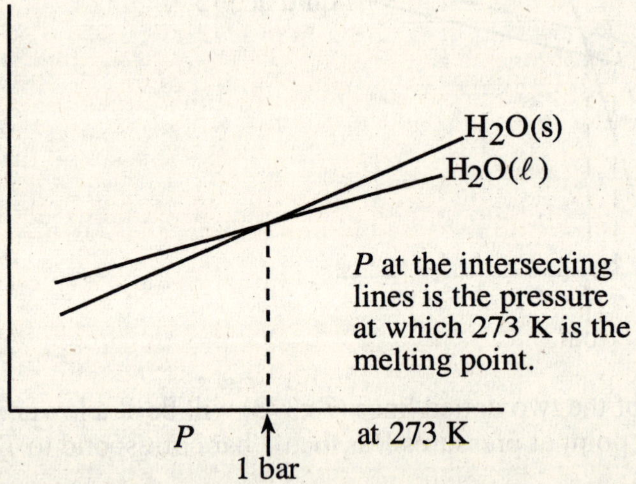

4.

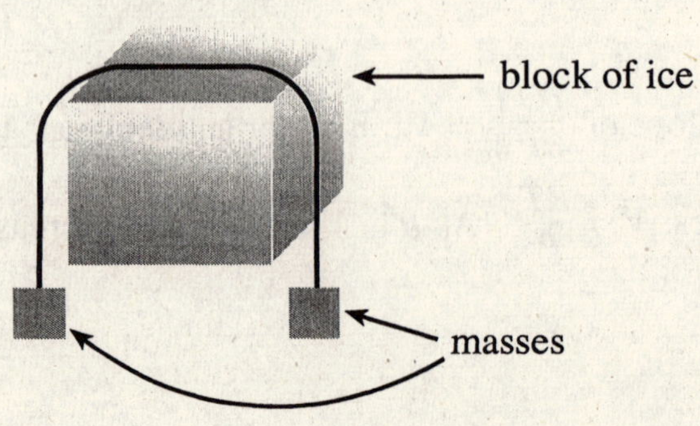

$$\left(\frac{\partial G}{\partial P}\right)_T = \overline{V}$$

When P is increased on the ice (at its freezing point), the value of $\overline{G}$ for the solid is increased more than for the liquid ($\overline{V}_{ice} > \overline{V}_{liquid}$ at freezing point).

Thus, $\overline{G}_{(liquid)} < \overline{G}_{(ice)}$ and so the ice melts. The wire then slips down into the ice, and continues this process all the way through.

5. $$\frac{d \ln P}{dT} = \frac{\Delta_r\overline{H}_{vap}}{RT^2} \qquad \int_{P_1}^{P_2} d \ln P = \int_{T_1}^{T_2} \frac{\Delta_r\overline{H}_{vap}}{RT^2}\, dT$$

$$\ln \frac{P_2}{P_1} = -\frac{\Delta_r\overline{H}_{vap}}{R}\left(\frac{1}{T_2} - \frac{1}{T_1}\right)$$ if $\Delta_r\overline{H}_{vap}$ is independent of T.

Thus, a plot of $\ln P$ vs $\frac{1}{T}$ would have a slope of $-\frac{\Delta_r\overline{H}_{vap}}{R}$ where P is the vapor pressure of the liquid at the temperature T.

6. $A(s) \rightleftharpoons A(g) \qquad \Delta_r\overline{H} = \Delta_r\overline{H}_{sub}$

$\overline{V}(s)dP - \overline{S}(s)dT = \overline{V}(g)dP - \overline{S}(g)dT$ at equilibrium

$$[\overline{V}(g) - \overline{V}(s)]dP = [\overline{S}(g) - \overline{S}(s)]\, dT$$

$$\frac{dP}{dT} = \frac{\Delta_r\overline{S}_{sub}}{\Delta_r\overline{V}} = \frac{\Delta_r\overline{H}_{sub}}{T\Delta_r\overline{V}}$$ at equilibrium ($\Delta_r\overline{H} = T\Delta_r\overline{S}$)

$$\frac{1}{P}\frac{dP}{dT} = \frac{d \ln P}{dT} = \frac{1}{P}\,\frac{\Delta_r\overline{H}_{sub}}{T\Delta_r\overline{V}}$$

Assume the gas is ideal and that $\overline{V}(s) << \overline{V}(g)$.

Then $\Delta_r\overline{V} = \overline{V}(g) - \overline{V}(s) = \overline{V}(g) = \frac{RT}{P}$.

Then $$\frac{d \ln P}{dT} = \frac{1}{P}\,\frac{\Delta_r\overline{H}_{sub}}{T\cdot\frac{RT}{P}} = \frac{\Delta_r\overline{H}_{sub}}{RT^2}$$

7. Analogous to Exercise 5, measure vapor pressure of the solid as a function of T.

Then a plot of $\ln P$ vs. $\frac{1}{T}$ would have a slope of $-\frac{\Delta_r \overline{H}_{sub}}{R}$.

8. $A(s) \rightleftharpoons A(\ell)$ $\qquad \Delta_r\overline{H} = \Delta_r \overline{H}_{fus}$

$\overline{V}(s)dP - \overline{S}(s)dT = \overline{V}(\ell)dP - \overline{S}(\ell)dT$ at equilibrium

$$\left[\overline{V}(\ell) - \overline{V}(g)\right]dP = \left[\overline{S}(\ell) - \overline{S}(s)\right] dT$$

$$\frac{dP}{dT} = \frac{\Delta_r \overline{S}_{fus}}{\Delta_r \overline{V}} = \frac{\Delta_r \overline{H}_{fus}}{T\Delta_r \overline{V}_{fus}} \text{ because } (\Delta_r\overline{H} = T\Delta_r \overline{S} \text{ at equilibrium})$$

9. $\frac{dP}{dT} = \frac{\Delta_r\overline{H}}{T\Delta_r \overline{V}}$ where $\Delta_r\overline{H} = H(\text{prod}) - H(\text{react})$ and $\Delta_r \overline{V} = V(\text{prod}) - V(\text{react})$.

$$\int_{P_1}^{P_2} dP = \frac{\Delta_r\overline{H}}{\Delta_r\overline{V}} \int_{T_1}^{T_2} \frac{dT}{T}$$ assuming $\Delta_r\overline{H}$, $\Delta_r \overline{V}$ are independent of T .

$$P_2 - P_1 = \frac{\Delta_r\overline{H}}{\Delta_r\overline{V}} (\ln T_2 - \ln T_1) = \frac{\Delta_r\overline{H}}{\Delta_r\overline{V}} \ln \frac{T_2}{T_1}$$

10. At $T_1 = 373$ K, $P_1 = 1$ bar. $\qquad T_2 = 353$ K; $\Delta_r\text{H} = 2255 \frac{\text{J}}{\text{g}} = 40{,}630$ J/mol

$$\ln \frac{P_2}{P_1} = -\frac{\Delta_r \overline{H}_{vap}}{R} \left(\frac{1}{T_2} - \frac{1}{T_1}\right)$$

$$\ln P_2 = -\frac{40{,}630 \text{ J/mol}}{8.315 \text{ J/Kmol}} \left(\frac{1}{353} - \frac{1}{373}\right)$$

$$= -(4.886 \times 10^3)(1.519 \times 10^{-4}) = -0.742$$

$\therefore P_2 = 0.476$ bar

11. $T_1 = 353.4$ K $\qquad P_1 = 1$ bar

$T_2 = 298.2$ K $\qquad P_2 = 0.13$ bar

$$\Delta_r\overline{H}_{vap} = \frac{R\ln\frac{P_2}{P_1}}{\left(\frac{1}{T_1}-\frac{1}{T_2}\right)} = \frac{\left(8.315\,\frac{\text{J}}{\text{Kmol}}\right)\ln(0.13)}{\left(\frac{1}{353.4}-\frac{1}{298.2}\right)}$$

= 32 kJ/mol

12. $$\frac{d\ln P}{dT} = \frac{\Delta_r\overline{H}^\circ_{vap}}{RT^2}$$

$$\frac{d\ln P}{dT} = \frac{2819.7}{T^2} + 1.855\times10^{-3} = \frac{\Delta_r\overline{H}^\circ_{vap}}{RT^2}$$

At 36.1°C $\Delta_r\overline{H}^\circ_{vap}$ = 24.9 kJ/mol .

13. Assume that the densities do not change with pressure.

Solid H_2O is 0.915 g/mL = 915 g/L = 50.8 $\frac{\text{mol}}{\text{L}}$

$$\therefore\ \overline{V}(s) = \frac{1}{50.8\text{ mol/L}} = 0.0197\,\frac{\text{L}}{\text{mol}}$$

Liquid H_2O is 1000 $\frac{\text{g}}{\text{L}}$ = 55.5 $\frac{\text{mol}}{\text{L}}$; $\overline{V}(\ell) = 0.0180\,\frac{\text{L}}{\text{mol}}$

$$\therefore\ \Delta_r\overline{V} = \overline{V}(\ell) - \overline{V}(s) = -0.0017\,\frac{\text{L}}{\text{mol}} = -0.17\times10^{-5}\ \text{m}^3/\text{mol}$$

$\Delta_r\overline{H}^\circ_{fus}$ = 6024 J/mol at T_1 = 273 K, P_1 = 1 bar, P_2 = 100 bar

$$P_2 - P_1 = \frac{\Delta_r\overline{H}}{\Delta_r\overline{V}}\ln\frac{T_2}{T_1}$$

Recall that $10^5\,\frac{\text{J}}{\text{m}^3}$ = 1 bar

$$100\text{ bar} - 1\text{ bar} = \frac{6024\text{ J/mol}}{-0.17\times10^{-5}\ \text{m}^3/\text{mol}}\ln\frac{T_2}{273}$$

$$99\text{ bar} = (-3.544\times10^4)\ln\frac{T_2}{273}$$

$$-2.79\times10^{-3} + \ln 273 = \ln T_2 = 5.607$$

$\therefore$ T_2 = 272.2 K (lower than 273, as expected)

14. $H_2O(s) \rightleftharpoons H_2O(g)$ at 273.15 °K .

ΔH for this process = $\Delta_r\overline{H}^\circ_{fus}$ (melting) + $\Delta_r\overline{H}^\circ_{vap}$ (vaporization) (Hess' Law)

$$\therefore\ \Delta_r\overline{H}^\circ_{sub} = +44.8\ \frac{kJ}{mol} + 6.024\ \frac{kJ}{mol} = +50.8\ \frac{kJ}{mol}$$

$$\frac{d\ln P}{dT} = \frac{\Delta_r\overline{H}_{sub}}{RT^2} \Rightarrow d\ln P = \frac{\Delta_r\overline{H}_{sub}}{R}\frac{dT}{T^2}$$

$$\int_{T=0°C}^{T=-15°C} d\ln P = \frac{\Delta_r\overline{H}_{sub}}{R}\int_{T=0°C=273\text{ K}}^{T=-15°C=258\text{ K}} \frac{dT}{T^2}$$ assuming $\Delta_r H_{sub}$ is

independent of T.

$$\ln P_{T=-15\ °C} - \ln P_{T=0\ °C} = \frac{+50{,}800\text{ J/mol}}{8.315\ \frac{J}{mol\ K}}\left(\frac{1}{273} - \frac{1}{258}\right)$$

$$\ln P_{T=258K} = \ln P_{T=273K} - 1.301$$

$$P_{273K} = \frac{4.58\text{ Torr}}{760\text{ Torr/atm}} \times 1.01325\ \frac{bar}{atm} = 6.1 \times 10^{-3}\text{ bar}$$

$$\ln P_{T=258K} = -5.098 - 1.301 = -6.399$$

$$P_{T=258K} = 1.66 \times 10^{-3}\text{ bar} = 1.25\text{ Torr}$$

15. $\left.\frac{\partial G}{\partial T}\right)_P = -\overline{S} \Rightarrow$ For solids, $\overline{S}$ is relatively constant as a function of T.

$d\overline{G} = \overline{V}dP - \overline{S}dT \qquad \overline{S}(C) > \overline{S}(A)$

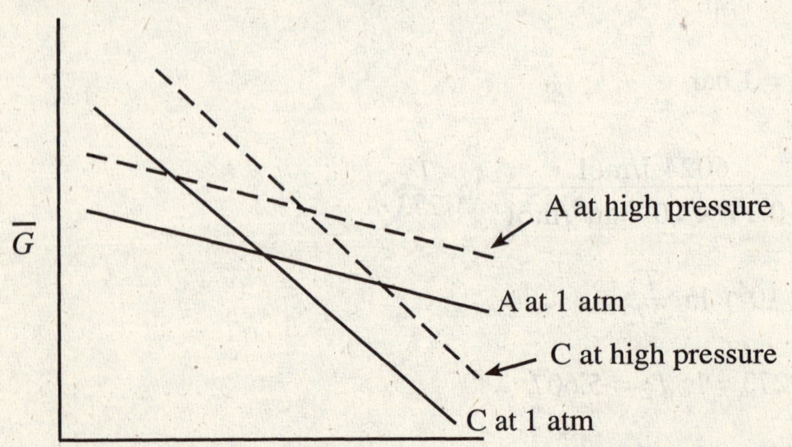

If $P\uparrow$, then $\overline{G}\uparrow$ because $\left.\frac{\partial \overline{G}}{\partial P}\right)_T = \overline{V}$ ·

Line for C is raised more (at high pressure) than line for A because $\overline{V}(\mathrm{C}) > \overline{V}(\mathrm{A})$.

Consider $\mathrm{A} \rightleftharpoons \mathrm{C}$

$\Delta_r H^\circ = 0.2 \frac{\text{kJ}}{\text{mol}}$, $\Delta_r S^\circ = 4.2 \frac{\text{J}}{\text{Kmol}}$

Assuming $\Delta_r H^\circ$, $\Delta_r S^\circ$ are independent of T:

At equilibrium: $\Delta_r G^\circ = 0 = \Delta_r H^\circ - T\Delta_r S^\circ$

$$0 = 200 \frac{\text{J}}{\text{mol}} - T\left(4.2 \frac{\text{J}}{\text{mol K}}\right) \Rightarrow T \doteq 48 \text{ K}$$

16. $\frac{dP}{dT} = \frac{\Delta \overline{S}}{\Delta \overline{V}} = \frac{-4.2}{-2.97} = 1.4 \times 10^1 \text{ atm K}^{-1}$

$P_2 - P_1 = 14\,(298–48) = P_2 - 1$

$P \approx 3500$ atm

$\Delta \overline{S}$ and $\Delta \overline{V}$ are assumed to be independent of T and P.

CA T14 Vapor Pressure and One Component Phase Diagrams

Exercises

1. At $T\downarrow$, vapor pressure $\downarrow$. Thus, some of the vapor condenses. Also, as T $\downarrow$, P $\downarrow$ according to the ideal gas law. So, both effects lead to a dramatic decrease in gas pressure. Air has been expelled from the flask by vigorous heating and the water will boil at all temperatures because the vapor pressure of water is the external pressure.

2. Recall that $\frac{dP}{dT} = \frac{\Delta_r\overline{H}}{T\Delta_r\overline{V}}$. $\Delta_r\overline{H}_{fus} < \Delta_r\overline{H}_{vap}$

   For solid $\rightleftharpoons$ liquid, $\Delta_r\overline{V} > 0$ but relatively small.

   For liquid $\rightleftharpoons$ gas, $\Delta_r\overline{V} > 0$ but large.

   For solid $\rightleftharpoons$ gas, $\Delta_r\overline{V} > 0$ but large.

   Assume $\Delta_r\overline{H}$, $\Delta_r\overline{V}$ (for solids, liquids) independent of T.

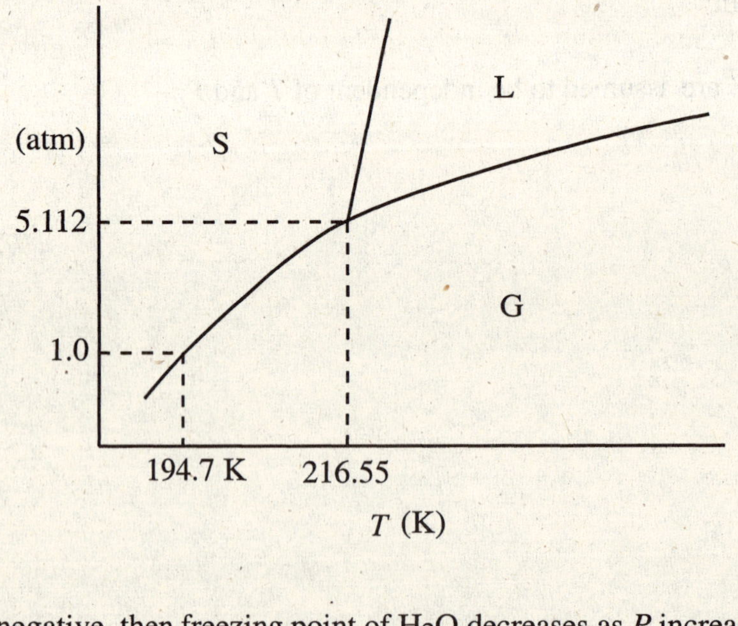

3. If slope is negative, then freezing point of H_2O decreases as P increases.

4. $\overline{V}gdP - \overline{S}gdT = \overline{V}\ell dP - \overline{S}\ell dT$

T is a constant.

$\overline{V}gdPg = \overline{V}\ell dP\ell$

$\frac{dPg}{dP\ell} = \frac{\overline{V}\ell}{\overline{V}g}$ The ratio $\frac{\overline{V}\ell}{\overline{V}g}$ is a positive number.

If the pressure on the liquid is increased, $dP\ell$, the vapor pressure, dPg, increases.

5. A - Solid H_2O at a very low temperature in equilibrium with a very low vapor pressure of $H_2O(g)$ at a total pressure of 1 bar.

B - As in A, but VP of $H_2O(g)$ is slightly higher, and T is higher.

C - Liquid H_2O (above 0 °C) with some vapor in equilibrium at a total pressure of 1 bar.

D - As in C, but VP of $H_2O(g)$ is higher, and T is higher.

E - $T = 100$ °C. Liquid is boiling. VP of gas = 1 bar.

F - $T > 100$ °C. Vapor is present at total pressure of 1 bar.

G - Solid is present at a low T. P is less than 1 bar. VP of gas present is less than P.

H - At this T, solid is present, and VP of gas is equal to total pressure, P.

I - Gas only is present at a total pressure of P.

CA T15 The Ideal Solution

Exercises

1. This mixture is not expected to be ideal because the predominant intermolecular force in H_2O is hydrogen bonding while that in dibutyl ether is dipole-dipole interaction. Because these types of interactions are different, the solution is not expected to be ideal.

2. Dalton's Law: $P_i = X_{i(vap)} \times P_{tot}$

   Raoult's Law: $P_i = X_{i(sol)} \times P_i^*$

   Both laws involve the partial pressure of a component i in a mixture of gases. Dalton's Law relates this partial pressure to the total pressure of <u>any</u> mixture of ideal gases. However, Raoult's Law is used to describe the partial pressure of a component <u>in equilibrium with</u> a mixture of liquids.

3. Dalton's Law may be used: $X_{bz(vap)} = \dfrac{P_{bz}}{P_{tot}}$

moles of *bz*	$X_{bz(vap)}$	$X_{tol(vap)}$
1.00	1.00	0
0.00	0	1.00
0.200	0.263	0.737
0.400	0.445	0.555
0.800	0.818	0.182
0.100	0.263	0.737
0.800	0.928	0.0721

4. $P^*_{tol} = 25$ Torr $P^*_{xy} = 5$ Torr

   $X_{tol(sol)} = 0.50$ $X_{xy(sol)} = 0.50$

   $P_{tol} = X_{tol(sol)} \times P^*_{tol} = (0.50)(25 \text{ Torr}) = 12.5 \text{ Torr}$

   $P_{xy} = X_{xy(sol)} \times P^*_{xy} = (0.50)(5 \text{ Torr}) = 2.5 \text{ Torr}$

   Composition is generally described in terms of mole fractions. Using Dalton's Law for the vapor phase:

$$X_{tol(vap)} = \frac{P_{tol(vap)}}{P_{tot}} = \frac{12.5 \text{ Torr}}{15 \text{ Torr}} = 0.83 \Rightarrow X_{xy} = 0.17$$

5. Assume that the mixture of two components, A and B, is ideal. Let $X_{A(sol)}$ be the mole fraction of A in solution. Then, from Raoult's Law,

$$P_A = X_{A(sol)} \times P_A^*$$

$$P_B = X_{B(sol)} \times P_B^*$$

but $X_{B(sol)} = 1 - X_{A(sol)}$, so $P_B = (1 - X_{A(sol)})P_B^*$

and $P_{tot} = X_{A(sol)} \times P_A^* + (1 - X_{A(sol)})P_B^*$

According to Dalton's Law, $X_{A(vap)} = \dfrac{P_A}{P_{tot}} = \dfrac{X_{A(sol)} \times P_A^*}{X_{A(sol)}P_A^* + (1-X_{A(sol)})P_B^*}$

Thus, $P_A^* = P_B^*$ is the condition under which liquid and vapor compositions are identical.

6. If $P_A^* = P_B^* = P^*$ then

$$X_{A(vap)} = \frac{X_{A(sol)}P^*}{X_{A(sol)}P^* - X_{A(sol)}\ P^* + P^*}$$

$$= \frac{X_{A(sol)}P^*}{P^*} = X_{A(sol)}$$

CA T16 Chemical Potential for a Component of a Solution

Exercises

1. The chemical potential of a component in a mixture depends on the Gibbs energy of the pure liquid component and another term which is a function of the mole fraction of that component in the mixture.

2. $\mu_{A(sol)} = \mu^*_{A(\ell)} + RT \ln X_{A(sol)}$

   Since $X_{A(sol)} < 1$ for a mixture, the second term is negative, and the Gibbs energy of A in solution is less than that of pure A.

   The same holds true for B.

3. Gibbs energy of A decreases.

4. The mole fraction of the solvent decreases (from 1) when a solute is added. Thus, Gibbs energy decreases as seen from the equation in Exercise 1.

5. When pure water is boiling, the Gibbs energy of the liquid phase equals the Gibbs energy of the vapor phase. When salt is then added to the liquid, the Gibbs energy of water (liquid) is decreased, but the Gibbs energy of the vapor phase is unchanged because salt is not volatile (does not vaporize). Thus, boiling stops because the liquid phase has the lower Gibbs energy. Recall that $\left(\frac{\partial G}{\partial T}\right)_P = -S$ and that $S_{(\ell)} < S_{(vap)}$. If T is raised, G is decreased for each phase. But G decreases more for vapor phase. Thus, at some higher T, G of both will become equal, as the $G_{(vap)}$ is decreasing faster. Thus, boiling occurs at a higher T.

CA T17 Partial Molar Quantities

Exercises

1. Assume that the tennis balls are much smaller than the bowling balls. Then, the tennis balls are able (for the most part) to fill in the spaces present between the bowling balls. Thus, on average, adding a tennis ball will not increase the total volume very much at all: $\left.\frac{\partial V}{\partial n_{TB}}\right)_{n_{BB}} \approx 0$. Thus, total volume is not affected by adding tennis balls.

2. The final volume could be the same as one of the components if the other component were small enough to fit into the spaces between (as in tennis balls and bowling balls).

3. $\mu_{A(sol)} = \mu^*_{A(\ell)} + RT \ln X_A$

$$\mu_{B(sol)} = \mu^*_{B(\ell)} + RT \ln X_B$$

$$\overline{V}_A = \left.\frac{\partial \mu_{A(sol)}}{\partial P}\right)_{T,n_A,n_B} = \left.\frac{\partial \mu^*_{A(\ell)}}{\partial P}\right) + \frac{\partial}{\partial P}[RT \ln X_A]$$

$$= \overline{V}_A{}^*{}_{(\ell)} + 0 = \overline{V}_A{}^*{}_{(\ell)} \text{ molar volume of pure A}$$

$$\overline{V}_{B(\ell)} = \overline{V}_B{}^*{}_{(\ell)} \text{ molar volume of pure B}$$

$$\Delta \overline{V}_{mix} = \overline{V}_{mix} - n_A \overline{V}_A{}^* - n_B V_B{}^* \qquad \text{but } \overline{V}_{mix} = n_A \overline{V}_A + n_B V_B$$

$$= n_A \overline{V}_A{}^* + n_B V_B{}^*$$

$\therefore \Delta \overline{V}_{mix} = 0$ for ideal solutions

$$\frac{\mu_A}{T} = \frac{\mu^*_{A(\ell)}}{T} + R \ln X_A$$

$$\left.\frac{\partial\left(\frac{\mu_A}{T}\right)}{\partial T}\right)_{P,n_A,n_B} = -\frac{\overline{H}_A}{T^2} = \frac{T\frac{\partial \mu^*_A}{\partial T} - \mu^*_A}{T^2} = \frac{-\overline{S}_A{}^* T - \mu^*_A}{T^2} = -\frac{\overline{H}_A{}^*}{T^2}$$

$\therefore \overline{H}_A = \overline{H}_A{}^* \leftarrow$ molar entropy of pure "A" Also $\Rightarrow \overline{H}_B = \overline{H}_B{}^*$

$\uparrow$

partial molar entropy
of "A" in ideal solution

Thus, as for volumes, $\Delta \overline{H}_{mix} = 0$.

$$\left(\frac{\partial \mu_A}{\partial T}\right)_{P,n_A,n_B} = -\overline{S}_A = -S_A^* + R \ln X_A$$

$$\text{and } \overline{S}_B = S_B^* - R \ln X_B$$

4. $\Delta \overline{G}_{mix} = G_{mix} - n_A \overline{G}_A - n_B \overline{G}_B$

$$= n_A\mu_A + n_B\mu_B - n_A\mu_A^* - n_B\mu_B^*$$

$$\uparrow \quad \uparrow$$

molar free energy of pure liquid

But $\mu_i = \mu_i^* + RT \ln X_i$, so

$$\Delta \overline{G}_{mix} = n_A(\mu_A^* + RT \ln X_A) + n_B(\mu_B^* + RT \ln X_B) - n_A\mu_A^* - n_B\mu_B^*$$

$$= n_A RT \ln X_A + n_B RT \ln X_B$$

$\Delta \overline{S}_{mix} = S_{mix} - n_A \overline{S}_A^* - n_B \overline{S}_B^*$ and from Exercise 3:

$$-\overline{S}_B = -S_B^* + R \ln X_B$$

$$\overline{S}_B = S_B^* - R \ln X_B$$

$$\Delta \overline{S}_{mix} = n_A (S_A^* - R \ln X_A) + n_B (S_B^* - R \ln X_B) - n_A S_A^* - n_B S_B^*$$

$$= -n_A R \ln X_A - n_B R \ln X_B$$

5. $\Delta \overline{H}_{mix} = 0$, $\Delta \overline{V}_{mix} = 0$ as shown above in Exercise 3.

CA T18 Colligative Properties

Exercises

1. $\Delta T_{bp} = 0.65\ \text{K}$ $\qquad K_{bp} = 5.03\ \text{K/molal}$

   $T_o = 349.90\ \text{K}$

   $$\therefore\ \text{m}_{\text{unknown}} = \frac{\Delta T_{bp}}{K_{bp}} = \frac{1.30\ \text{K}}{5.03\ \text{K/molal}} = 0.258\ \frac{\text{moles}}{1000\ \text{g solvent}}$$

   $\therefore\ 1.29 \times 10^{-2}$ moles dissolved in 50.00 g CCl_4

   $$\therefore\ \frac{1.24\ \text{g}}{1.29 \times 10^{-2}\ \text{mol}} = \text{MW} = 96.1\ \text{g/mole}$$

2. This derivation is analogous to the development in CTQs 9-11.

   $$\int_{X=1}^{X_{A(sol)}} d \ln X_{A(sol)} = \frac{\Delta_r \overline{H}_{\text{fus}}}{R} \int_{T_o}^{T} \frac{dT}{T^2}$$

   $$\ln X_{A(sol)} = \frac{-\Delta_r \overline{H}_{\text{fus}}}{R} \left(\frac{1}{T} - \frac{1}{T_o}\right)$$

   $$\ln X_{A(sol)} = \frac{-\Delta_r \overline{H}_{\text{fus}}}{R} \left(\frac{T_o - T}{T \cdot T_o}\right) \approx \frac{-\Delta_r H_{\text{fus}}}{R} \frac{\Delta T_{fp}}{T_o^2}$$

   when $T_o - T$ is a small number (ΔT_{fp} is small)

   $$\text{Let m} = \frac{\text{moles B}}{1000\ \text{g solvent A}}$$

   $$\ln X_A = \ln(1 - X_B) \approx -X_B = \frac{-\overline{\Delta_r H}_{\text{fus}}}{R} \frac{\Delta T_{fp}}{T_o^2}$$

   (arrow pointing to $\approx -X_B$: from Taylor series)

   For a dilute solution of B in A
   ↓
   $$\therefore\ X_B \approx \frac{\text{moles B}}{\text{moles A}} = \frac{\Delta_r H_{\text{fus}}}{R} \times \frac{\Delta T_{fp}}{T_o^2}$$

   $$\Delta T_{fp} = \frac{R T_o^2}{\Delta_r H_{\text{fus}}} \times \frac{\text{moles B}}{\text{moles A}} \quad \text{but moles A} = \frac{\text{kg A}}{\text{MW}_A} \leftarrow \text{in kg/mol}$$

$$\therefore\ \Delta T_{fp} = \frac{RT_o^2}{\Delta_r H_{fus}} \ \frac{\text{moles B}\cdot \text{MW}_A}{\text{kg A}} = \frac{RT_o^2 \cdot \text{MW}_A}{\Delta_r H_{fus}} \times \frac{\text{moles B}}{\text{kg A}}$$

$$\Delta T_{fp} = K_{fp} \times m_B \ \text{ where } K_{fp} = \frac{RT_o^2 \cdot \text{MW}_A}{\Delta_r H_{fus}}$$

3. $A(s) \rightleftharpoons A(\ell)$ $\qquad \Delta H^\circ = \Delta_r \overline{H}_{fus}$

$\underline{A(\ell) \rightleftharpoons A(sol) \qquad \Delta H^\circ = 0}$ $\qquad$ because $\Delta \overline{H}_{mix}$ for an ideal solution = 0

$\therefore$ $A(s) \rightleftharpoons A(sol)$ $\qquad \Delta H^\circ = \Delta_r \overline{H}_{fus}$ $\quad$ for an ideal solution

4. $$\ln X_{I_2(sol)} = \frac{\Delta_r \overline{H}_{fus}}{R} \left(\frac{T - T_o}{T \cdot T_o} \right)$$

$T = 298$ K $\qquad \Delta_r \overline{H}_{fus} = 15{,}640$ J/mol

$T_o = 386$ K $\qquad R = 8.315$ J/mol K

$\ln X_{I_2(sol)} = -1.439 \Rightarrow X_{I_2} = 0.237$ for an ideal solution

5. The solubility of "B" in "A" at a temperature T is the mole fraction of B which gives rise to a freezing point depression of $\Delta T_{fp} = T_o - T$.

CA T19 Osmotic Pressure

Exercises

1. Assuming that 1.0 M sucrose is dilute enough for the derivation in CA T19 to apply:

$$\pi = R \cdot T \cdot C_{suc}$$

$$= \left(8.315 \times 10^{-2} \frac{\text{L·bar}}{\text{mol K}}\right) (298 \text{ K}) \left(1.0 \frac{\text{mol}}{\text{L}}\right)$$

$$= 25 \text{ bar}$$

$$\text{or } \pi = \left(.08206 \frac{\text{L·atm}}{\text{mol K}}\right) (298 \text{ K}) \left(1.0 \frac{\text{mol}}{\text{L}}\right) \times 760 \frac{\text{Torr}}{\text{atm}}$$

$$= 1.9 \times 10^4 \text{ Torr} = 1.9 \times 10^4 \text{ mm Hg}$$

Density of Hg is 13.6 g/ml; density of H_2O = 1.0 g/mol.

∴ column of H_2O would be 13.6/1.0 = 136 times as high to exert the same pressure.

∴ column of H_2O would be (13.6) (1.9×10^4 mm) = 2.5×10^5 mm H_2O

$$= 2.5 \times 10^2 \text{ m}$$

CA T20 The Phase Rule

Exercises

1. None ⇒ system is completely specified (assuming 2 phases).

2. Place H_2O in an otherwise evacuated container with a pressure meter attached. Vary the temperature of the system until the pressure reads 23.8 Torr.

   No other variables are needed to be known - there is only one *T* for which vapor pressure of pure H_2O is 23.8 Torr.

   The only intensive variables are *P*, *T*.

3. For sucrose in water, there are 3 intensive variables. If we assume that the solution to be prepared is a saturated solution in equilibrium with $H_2O(g)$ at its vapor pressure, then <u>one</u> of the three intensive variables (T, P, X_{suc}) must be specified.

   For example, if *T* were specified, this would determine the maximum solubility of sucrose. Once the solubility of sucrose was determined, the vapor pressure of H_2O would be determined through Raoult's Law (assuming ideal behavior).

4. If $c = 1$, and $f = 0$ (minimum possible),

   then $p = c - f + 2$
   $= 1 - 0 + 2$
   $= 3$

   ∴ at most 3 phases for a 1-component system.

   If $c = 2$, $p_{max} = 4$

   If $c = 3$, $p_{max} = 5$

5. 3 phases, $K = P_{CO_2}$, two components.

6. 2 components, 1 phase

   Any combination of these three gases at equilibrium could be produced by combining (originally) only 2 of these species. Another way of saying this is that the third species can always be produced from the other two (related through the equilibrium constant), so there are only 2 chemically-independent species. If it is specified that PCl_3 and Cl_2 are present in equimolar amounts, only one component is needed.

CA T21 Solid-Liquid Phase Equilibria

Exercises

1. a) For this problem, the eutectic mixture actually occurs at $X_{Pb} = 0.855$.

At this composition, the "melting temperature" of the "impure" Pb (X_{Pb} = 0.855) is equal to the melting temperature of the "impure" Sb (X_{Sb} = 0.145).

If the enthalpies of fusion are known, the eutectic temperature can be calculated. Let T be this temperature. Then both of the relationships below must be true.

$$\ln X_{Pb} = -\frac{\Delta_r\overline{H}_{fus(Pb)}}{R}\left(\frac{1}{T} - \frac{1}{T_{o(Pb)}}\right)$$

$$\ln X_{Sb} = -\frac{\Delta_r\overline{H}_{fus(Sb)}}{R}\left(\frac{1}{T} - \frac{1}{T_{o(Sb)}}\right)$$

so $$\ln(0.855) = -\frac{5{,}100\text{ J/mol}}{8.315\text{ J/Kmol}}\left(\frac{1}{T} - \frac{1}{600.55\text{ K}}\right)$$

$\Rightarrow T = 521$ K

$$\ln(0.145) = -\frac{20{,}100\text{ J/mol}}{8.315\text{ J/Kmol}}\left(\frac{1}{T} - \frac{1}{903.7\text{ K}}\right)$$

$\Rightarrow T = 525$ K

These two values of T are very close, so we conclude that this is a eutectic mixture, with $T_{eutectic} \approx 523\text{ K} \approx 250\text{ °C}$.

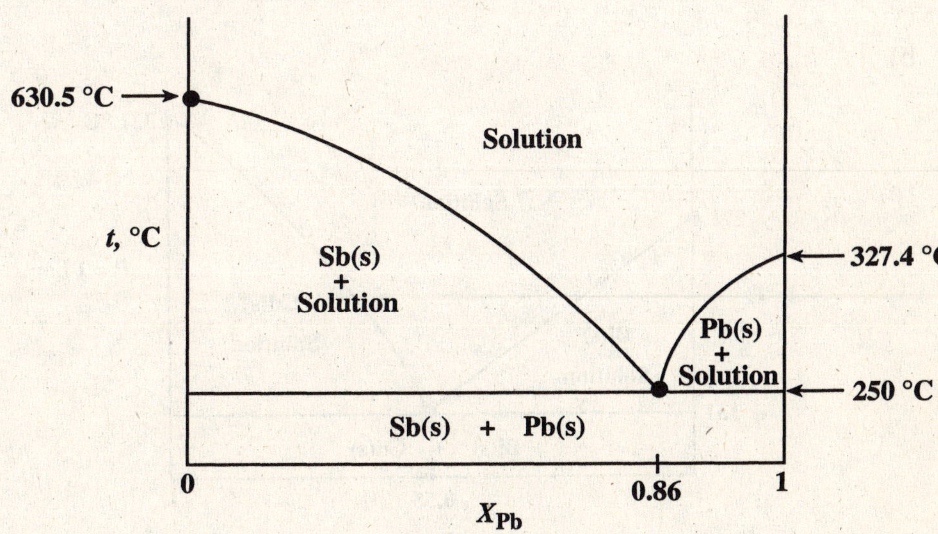

b) At eutectic point $f = 2 - 3 + 2 = 1$
<u>but since at 1 bar pressure</u>, then P is fixed, so no (zero) degrees of freedom

In solution: $f = 2 - 1 + 2 = 3$
but only 2 since at 1 bar pressure

Sb(s) + solution or Pb(s) + solution: $f = 2 - 2 + 2 = 2$
but only 1 since at 1 bar pressure

Sb(s) + Pb(s): $f = 2 - 2 + 2 = 2$
but only 1 since at 1 bar pressure

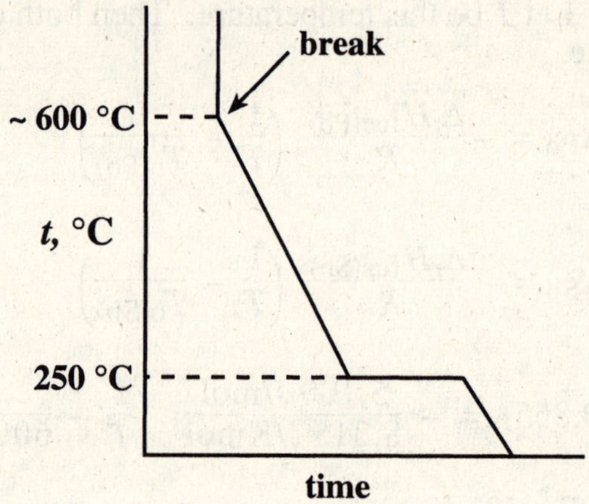

2. Proceed as in Exercise 1.

a) $$\ln X_{Cd} = -\frac{6.19 \times 10^3}{8.314}\left(\frac{1}{T_{eu}} - \frac{1}{594}\right), \ \ln X_{Bi} = \frac{-11.30 \times 10^3}{8.314}\left(\frac{1}{T_{eu}} - \frac{1}{544}\right)$$

$$T_{eu} = 404 \text{ K} = 131 \text{ °C}$$

b)

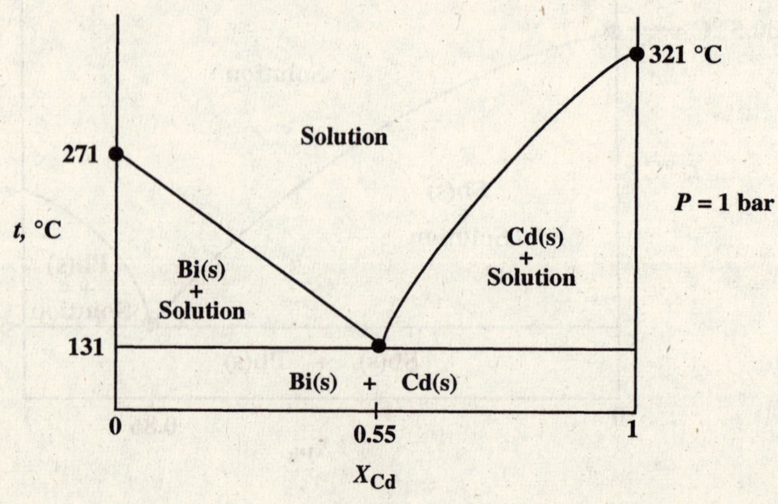

c)

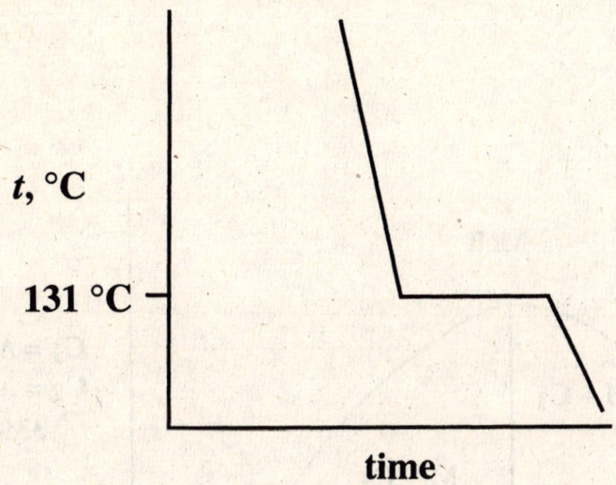

CA T22 More Complicated Solid-Liquid Phase Equilibria

Exercises

1.

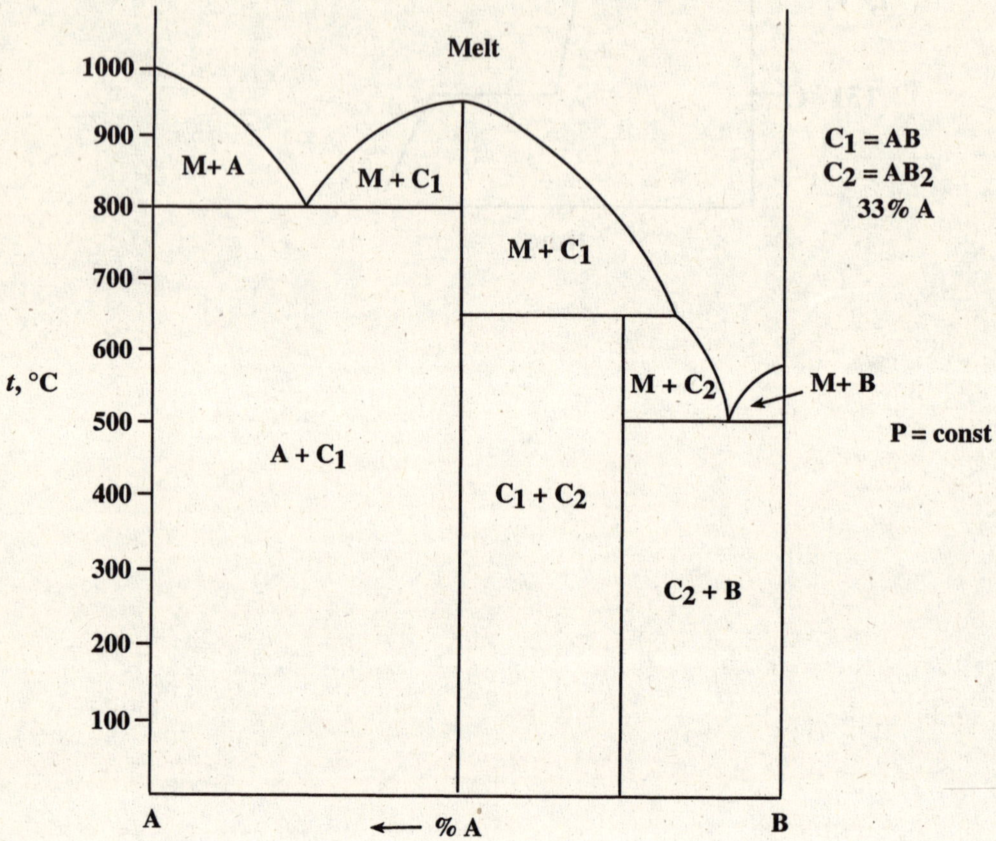

Note that at a peritectic point there are three phases. $f = c - p + 1 = 2 - 3 + 1 = 0$ but that the system must be treated as one of two components because the components are in different phases.

$$AB_2(s) = AB(s) + AB(melt)$$

In order to prepare arbitrary amounts of all phases, two components are needed. Contrast this to a congruent melting point.

$AB(s) = AB(melt)$ only one component is necessary and $f = c - p + 1$
$f = 1 - 2 + 1 = 0.$

2. C = NaK

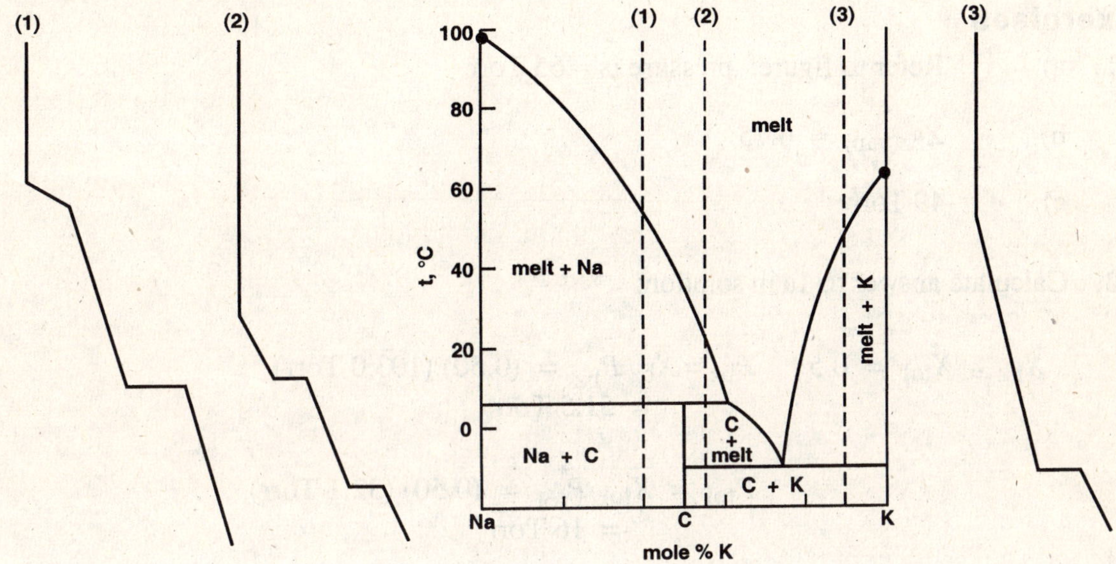

3.

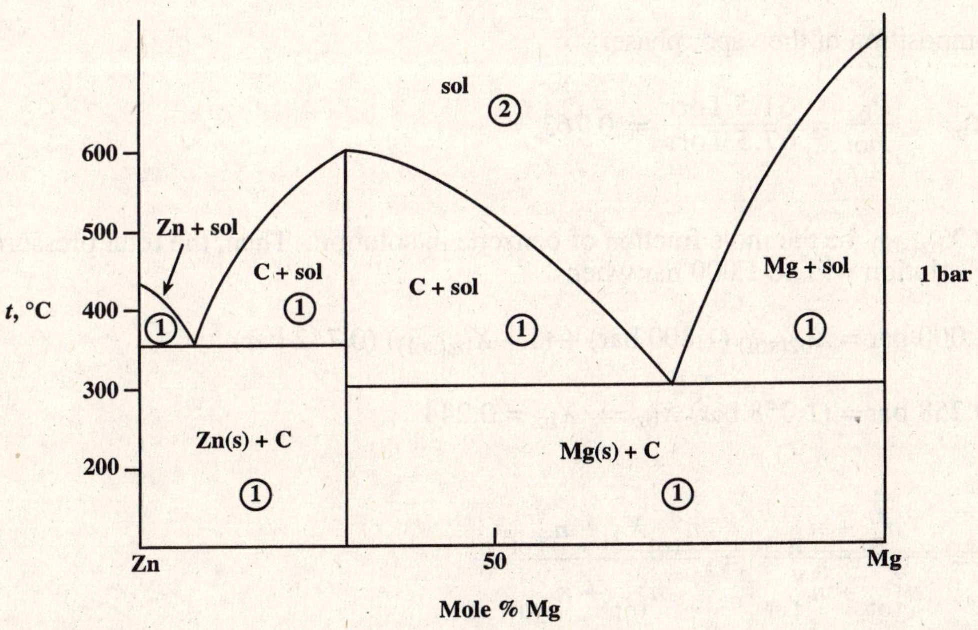

$C = MgZn_2$

$f = c - p + 1$

at eutectics $f = 0$

0 = degrees of freedom

CA T23 Liquid-Vapor Phase Equilibria

Exercises

1. a) Refer to figure: pressure is ≈ 65 Torr

   b) $X_{bz(vap)} \approx 0.75$

   c) 49 Torr

2. Calculate answer to 1a in solution:

$X_{bz} = X_{tol} = 0.5$ $\quad P_{bz} = X_{bz} P^*_{bz} = (0.50)\ (103.0\ \text{Torr}) = 51.5\ \text{Torr}$

$P_{tol} = X_{tol} P^*_{tol} = (0.50)\ (32.1\ \text{Torr}) = 16\ \text{Torr}$

$P_{tot} = P_{bz} + P_{tol} = 67.5\ \text{Torr}$

At this external pressure, solution will "boil."

Composition of the vapor phase:

$$X_{bz} = \frac{P_{bz}}{P_{tot}} = \frac{51.5\ \text{Torr}}{67.5\ \text{Torr}} = 0.763$$

3. Let $X_{bz(sol)}$ be the mole fraction of benzene in solution. Then, the total pressure of the solution will be 1.000 bar when

$$1.000\ \text{bar} = X_{bz(sol)}\ (1.800\ \text{bar}) + (1 - X_{bz(sol)})\ (0.742\ \text{bar})$$

$$0.258\ \text{bar} = (1.058\ \text{bar})\ X_{bz} \Rightarrow X_{bz} = 0.244$$

4. $$X_2 = \frac{n^{\ell}_{B} + n^{v}_{B}}{n^{\ell}_{tot} + n^{v}_{tot}} = \frac{n^{\ell}_{tot}X_1 + n^{v}_{tot}X_3}{n^{\ell}_{tot} + n^{v}_{tot}}$$

$$n^{\ell}_{tot}X_2 + n^{v}_{tot}X_2 = n^{\ell}_{tot}X_1 + n^{v}_{tot}X_3$$

$$n^{\ell}_{tot}(X_2 - X_1) = n^{v}_{tot}(X_3 - X_2)$$

$$\frac{n^{\ell}_{tot}}{n^{v}_{tot}} = \frac{X_3 - X_2}{X_2 - X_1} = \frac{DE}{DC}$$

5. $X_{bz(vap)} = 0.64$, $X_{bz(sol)} = 0.36$

6. a) $X_c \approx 0.65$ $X_a \approx 0.35$
for both liquid and vapor

b) Not possible. Starting with a 50-50 mixture, it is possible to obtain pure acetone through a series of sequential distillations (redistilling the condensed distillate).

However, the highest mole fraction of chloroform which can be obtained is $X_c = 0.65$ (the azeotropic composite).

c)

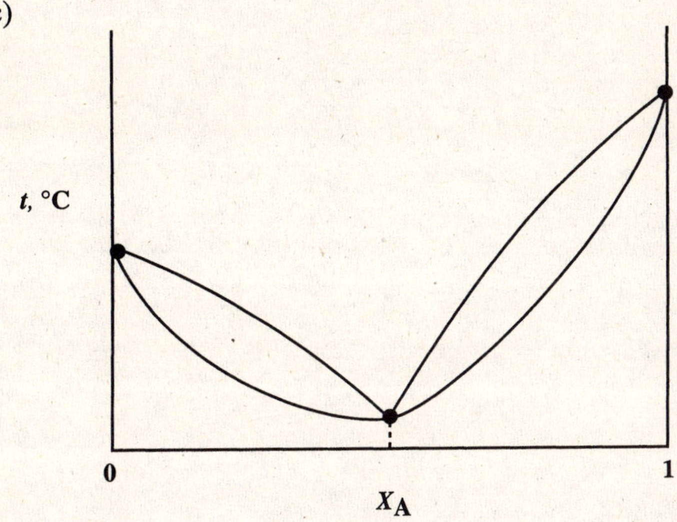

This mixture exhibits a positive deviation from Raoult's Law.

The interactions in the mixture are weaker than in the pure liquids.

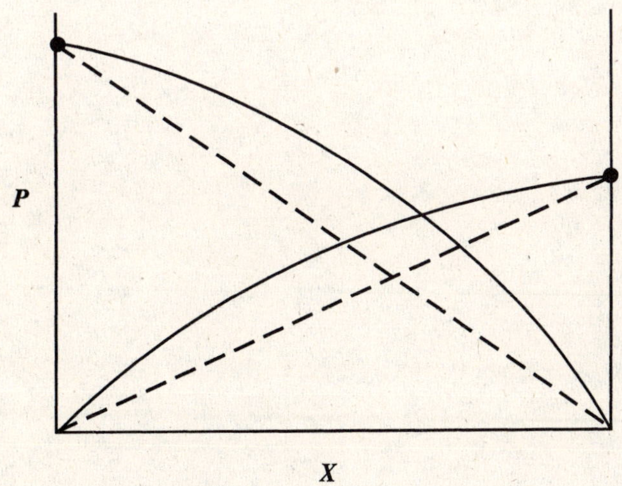

Observed partial pressures are higher than expected from Raoult's Law.

d) Consider minimum boiling azeotroic mixture: interactions in mixture are stronger than in pure liquids.

Thus, $\Delta\overline{H}_{mix} < 0$; $\Delta\overline{V}_{mix} < 0$

$\Delta\overline{H}_{mix} < 0$ ↑ stronger interactions means more negative enthalpy

$\Delta\overline{V}_{mix} < 0$ ↑ stronger interactions means decreased volume

CA E1 Electrolyte Solutions

Exercises

1. $m_\pm^\nu = m_+^{\nu_+} m_-^{\nu_-}$, $m_+ = \nu_+ m$, $m_- = \nu_- m$

$$m_\pm^\nu = (\nu_+^{\nu_+} m^{\nu_+})(\nu_-^{\nu_-} m^{\nu_-})$$
$$= (\nu_+^{\nu_+} \nu_-^{\nu_-})(m^{\nu_+ + \nu_-}) = (\nu_+^{\nu_+} \nu_-^{\nu_-})\, m^\nu$$
$$m_\pm = (\nu_+^{\nu_+} \nu_-^{\nu_-})^{1/\nu}\, m$$

2. $m_\pm = (\nu_+^{\nu_+} \nu_-^{\nu_-})^{1/\nu}\, m$

$NaCl$	$m_\pm = m$
CaF_2	$m_\pm = (1^1\, 2^2)^{1/3}\, m = 1.59\, m$
Li_2SO_4	$m_\pm = (2^2\, 1^1)^{1/3}\, m = 1.59\, m$

3.

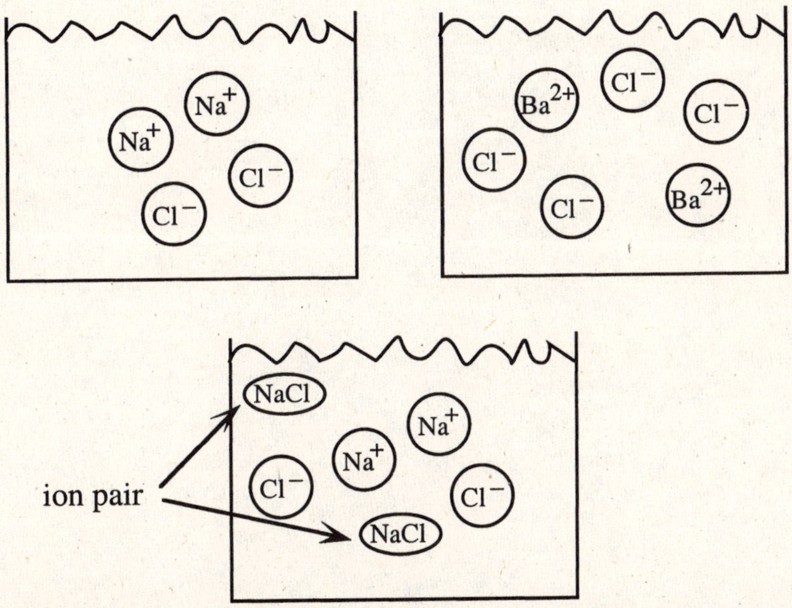

In highly concentrated solutions ion pairs can form. The greater the charge on the ions the more likely the formation of ion pairs.

4. Electrolyte solutions exhibit very strong interactions between ions and the solvent and between the ions themselves. These interactions are clearly stronger than typical solvent-solvent interactions, and they may extend over greater distances. One result is that ions are generally not uniformly distributed throughout a solution, but tend to form ion "clouds" around an oppositely-charged species. Thus, ideal behavior is not generally observed.

5. $0.03\ \mathrm{mol/(kg\ H_2O)}\ Ag_2SO_4$ $\quad m = 0.03$

$\nu_+ = 2;\quad \nu_- = 1;\quad \nu = 3$

conc. $Ag^+ = 0.06\ m$, $\quad$ conc. $SO_4^{2-} = 0.03\ m$

$m_\pm = (4^{1/3})(0.03\ m) = 0.05\ m$

$1.0 \times 10^{-3}\ \mathrm{mol/(kg\ H_2O)}\ PbF_2$ $\quad m = 1.0 \times 10^{-3}$

$\nu_+ = 1;\quad \nu_- = 2;\quad \nu = 3$

conc. $Pb^{2+} = 1.0 \times 10^{-3}\ m$, $\quad$ conc. $F^- = 2.0 \times 10^{-3}\ m$

$m_\pm = (4^{1/3})(1.0 \times 10^{-3}\ m) = 1.6 \times 10^{-3}\ m$

$10^{-4}\ \mathrm{mol/(kg\ H_2O)}\ BaC_2O_4$ $\quad m = 1 \times 10^{-4}$

$\nu_+ = 1;\quad \nu_- = 1;\quad \nu = 2$

conc. $Ba^{2+} = 1 \times 10^{-4}\ m$, $\quad$ conc. $C_2O_4^{2-} = 1 \times 10^{-4}\ m$

$m_\pm = (1)(1 \times 10^{-4}\ m) = 1 \times 10^{-4}\ m$

CA E2 The Debye-Hückel Theory of Electrolyte Solutions

Exercises

1. $e^{-x} \approx 1 - x$

$$\rho = e\left\{Z_+ n_+^0\left(1 - \frac{eZ_+ V}{kT}\right) + Z_- n_-^0\left(1 - \frac{eZ_- V}{kT}\right)\right\}$$

$$= e\left\{Z_+ n_+^0 - \frac{eZ_+^2 n_+^0 V}{kT} + Z_- n_-^0 - \frac{eZ_-^2 n_-^0 V}{kT}\right\}$$

$$= \frac{-e^2 V}{kT}\left(n_+^0 Z_+^2 + n_-^0 Z_-^2\right)$$

2. As r gets very large, V approaches infinity. This is an unrealistic solution.

3. a) $RT \ln \gamma_- = -\dfrac{(Z_- e)^2 \kappa}{2\varepsilon}$

$$\ln \gamma_- = -\frac{(Z_- e)^2 \kappa}{2\varepsilon RT} = -A\, Z_-^2\, I^{1/2}$$ see CTQ 11.

b) $\gamma_\pm^\nu = \gamma_+^{\nu_+}\, \gamma_-^{\nu_-}$

$$\nu \ln \gamma_\pm = \nu_+ \ln \gamma_+ + \nu_- \ln \gamma_-$$

From CTQ 11,

$$\nu \ln \gamma_\pm = \nu_+ \frac{(-Z_+ e)^2 \kappa}{2\varepsilon kT} + \nu_- \frac{(-Z_- e)^2 \kappa}{2\varepsilon kT}$$

$$\nu \ln \gamma_\pm = -\frac{e^2 \kappa}{2\varepsilon kT}\left(\nu_+ Z_+^2 + \nu_- Z_-^2\right)$$

$$\nu_+ Z_+ + \nu_- Z_- = 0$$

or $\nu_+ Z_+ = -\nu_- Z_-$, $\nu_+ Z_+^2 = -\nu_- Z_+ Z_-$

and $\nu_- Z_-^2 = -\nu_+ Z_+ Z_-$

$$\nu_+ Z_+^2 + \nu_- Z_-^2 = -(\nu_+ + \nu_-) Z_+ Z_- = -\nu Z_+ Z_-$$

$$\nu \ln \gamma_\pm = -\frac{e^2 \kappa}{2\varepsilon kT}(-\nu Z_+ Z_-)$$

$$\ln \gamma_\pm = A\, Z_+ Z_-\, I^{1/2}$$

4. $I = \frac{1}{2} \sum_i m_i Z_i^2$

$NaCl \quad I = \frac{1}{2} [0.050\,(1)^2 + 0.050\,(-1)^2] = 0.050$

$CaCl_2 \quad I = \frac{1}{2} [0.050\,(2)^2 + 0.050\,(-1)^2] = 0.15$

$MgSO_4 \quad I = \frac{1}{2} [0.050\,(2)^2 + 0.050\,(-2)^2] = 0.20$

CA E3 The Structure of Electrolyte Solutions

Exercises

1. $\ln \gamma_{\pm} = 1.17\, Z_+ Z_- I^{1/2}$, $I = 0.0150$, $\gamma_{\pm} = = 0.751$

2. $\ln \gamma_{\pm} = 1.17\, Z_+ Z_- I^{1/2}$,

HCl	1.0×10^{-4} m ,	$I = 1.0 \times 10^{-4}$ m , $\gamma_{\pm} = 0.99$
	1.0×10^{-3} m ,	$I = 1.0 \times 10^{-3}$ m , $\gamma_{\pm} = 0.96$
$CuSO_4$	1.0×10^{-4} m ,	$I = 4.0 \times 10^{-4}$ m , $\gamma_{\pm} = 0.91$
	1.0×10^{-3} m ,	$I = 4.0 \times 10^{-3}$ m , $\gamma_{\pm} = 0.74$

As ionic strength increases $\gamma_{\pm}$ decreases for both solutions. But, for the more highly charged Cu^{2+} and SO_4^{2-} at the same concentrations as HCl $\gamma_{\pm}$ is smaller. Larger charges are more stabilized in an ionic atmosphere.

3. $HA(aq) = H^+(aq) + A^-(aq)$

$$K = \frac{m_{H^+}\, m_{A^-}}{m_{HA}} \quad \text{assuming ideality}$$

$$K = \frac{\alpha^2 m_o^2}{m_o(1-\alpha)} = \frac{\left(2.91\times10^{-3}\right)^2}{(0.00749)} = 1.13 \times 10^{-3}$$

If nonideality is assumed, $K = \dfrac{\gamma_{\pm}^2\, m_{H^+}\, m_{A^-}}{m_{HA}}$

$I = 2.91 \times 10^{-3}$

$\ln \gamma_{\pm} = -1.17\, I^{1/2}$, $\gamma_{\pm} = 0.939$

$$K = \frac{(0.939)^2\left(2.91\times10^{-3}\right)^2}{(0.00749)} = 9.96 \times 10^{-4}$$

γ for HA is taken to be unity because HA has no charge.

4. $K = \dfrac{\left(m_{H^+}\right)^2}{m_o(1-\alpha)}$ neglect α and $\gamma_{\pm}$

$m_{H^+} = 1.3 \times 10^{-3}$

If $m_{H^+} = 1.3 \times 10^{-3}$, $I = 1.3 \times 10^{-3}$, and $\gamma_{\pm} = 0.96$

A further iteration is necessary.

$$K = \frac{\gamma_{\pm}^2 \left(m_{H^+}\right)^2}{m_o(1-\alpha)} \quad \text{again, neglect } \alpha$$

$$m_{H^+} = 1.4 \times 10^{-3}$$

$I = 1.4 \times 10^{-3}$, and $\gamma_{\pm} = 0.96$ no further iterations are needed.

5. $K_{sp} = [a_+][a_-] = \gamma_{\pm}^2 S^2$

$$\ln K = 2 \ln \gamma_{\pm} + 2 \ln S$$
$$= -2\,(1.17)\, I^{1/2} + 2 \ln S$$

$$\frac{\ln K}{2} = \ln S - 1.17\, I^{1/2}$$

at zero ionic strength, $\dfrac{\ln K}{2} = \ln S_o$

$$\ln S_o = \ln S + \ln \gamma_{\pm}$$

$$\frac{S_o}{S} = \gamma_{\pm}$$

6. The solubility increases.

7. a) $K_{sp} = a_{Pb^{2+}}\, a_{I^-}^2 = \gamma_+\, m_{Pb^{2+}}\, \gamma_-^2\, m_{I^-}^2$

$$= \gamma_{\pm}^3\,(S)\,(2S)^2 = \gamma_{\pm}^3\, 4S^3$$

$$I = \frac{1}{2}\left[1.37 \times 10^{-3}\,(2)^2 + (2 \times 1.37 \times 10^{-3})(-1)^2\right]$$
$$= 4.11 \times 10^{-3}$$

$$\ln \gamma_{\pm} = 1.17\,(+2)\,(-1)\,(4.11 \times 10^{-3})^{1/2}$$
$$\gamma_{\pm} = 0.861$$

$$K_{sp} = (0.861)^3\, 4\,(1.37 \times 10^{-3})^3 = 6.52 \times 10^{-9}$$

b) $I = \dfrac{1}{2}\left[0.030\,(1)^2 + 0.030(-1)^2\right] = 0.030$

$$\ln \gamma_{\pm} = 1.17\,(+2)\,(-1)\,(0.030)^{1/2}$$
$$\gamma_{\pm} = 0.67$$

$$K_{sp} = 6.52 \times 10^{-9} = \gamma_{\pm}^3\, m_{Pb^{2+}}\, m_{I^-}^2 = (0.67)^3\, S\, (0.030)^2$$

$$S = 2.4 \times 10^{-5}\ \text{m}$$

c) Iteration is needed. First, assume $\gamma_\pm = 0.67$, $I = 0.030$, as in (b).

$$6.52 \times 10^{-9} = \gamma_\pm^3 \, m_{Pb^{2+}} \, m_{I^-}^2 = \gamma_\pm^3 \, 4S^3 \ , \quad S = 1.74 \times 10^{-3}$$

$$\text{New I} = 0.030 + \frac{1}{2}\left[1.74 \times 10^{-3} (2)^2 + 2(1.74 \times 10^{-3})(-1)^2\right] = 0.035$$

New $\gamma_\pm = 0.65$

$$K_{sp} = \gamma_\pm^3 \, 4S^3 = (0.65)^3 \, 4S^3 \qquad S = 1.8 \times 10^{-3}$$

New I ≈ 0.036 New $\gamma_\pm = 0.64$ New S $= 1.8 \times 10^{-3}$

d) $K_{sp} = 6.52 \times 10^{-9}$ for PbI_2 at 293 K. If $\gamma_\pm$ is ignored,

$$K_{sp} = \gamma_\pm^3 \, 4S^3 = 4S^3 \ , \quad S = 1.18 \times 10^{-3}$$

The experimental solubility is 1.37×10^{-3}, a 14% difference.

CA E4 Electrode Potentials

Exercises

1. At equilibrium: $\mu_{M^{n+}(aq)} + n\,\overline{\mu}_{e(M)} = \mu_{M(s)}$

$$\mu^{\theta}_{M^{n+}(aq)} + RT\ln a_{M^{n+}(aq)} + n(-1)F\,V = \mu^{*}_{M(s)}$$

$$V = \frac{\mu^{*}_{M(s)} - \mu^{\theta}_{M^{n+}(aq)} - RT\ln a_{M^{n+}(aq)}}{-nF}\,,\ \mu^{*}_{M(s)} \to 0$$

$$V = \frac{\mu^{\theta}_{M^{n+}(aq)} + RT\ln a_{M^{n+}(aq)}}{nF}$$

2. At equilibrium: $\mu_{Cl_2(g)} + 2\,\overline{\mu}_{e(Pt)} = 2\,\mu_{Cl^-(aq)}$ (assumes Pt electrode)

$$\mu^{\theta}_{Cl_2(g)} + RT\ln P_{Cl_2(g)} + 2(-1)F\,V = 2\,\mu^{\theta}_{Cl^-(aq)} + 2RT\ln a_{Cl^-(aq)}$$

$$\mu^{\theta}_{Cl_2(g)} \to 0$$

$$V = \frac{2\mu^{\theta}_{Cl^-(aq)} + 2RT\ln a_{Cl^-(aq)} - RT\ln P_{Cl_2(g)}}{-2F}$$

$$V = -\frac{\mu^{\theta}_{Cl^-(aq)}}{F} - \frac{RT\ln\dfrac{a^2_{Cl^-(aq)}}{P_{Cl_2(g)}}}{2F}$$

3. $\mu^{\theta}_{Fe^{3+}(aq)} + RT\ln a_{Fe^{3+}(aq)} + \overline{\mu}_{e(Fe)} = \mu^{\theta}_{Fe^{2+}(aq)} + RT\ln a_{Fe^{2+}(aq)}$

$$\mu^{\theta}_{Fe^{3+}(aq)} + RT\ln a_{Fe^{3+}(aq)} + (-1)F\,V = \mu^{\theta}_{Fe^{2+}(aq)} + RT\ln a_{Fe^{2+}(aq)}$$

$$V = \frac{\mu^{\theta}_{Fe^{2+}(aq)} - \mu^{\theta}_{Fe^{3+}(aq)} + RT\ln\dfrac{a_{Fe^{2+}(aq)}}{a_{Fe^{3+}(aq)}}}{-F}$$

4. By convention, electrode potentials are defined to be potentials associated with reduction processes. Thus, electrons are considered to be "reactants", resulting in reduction for each process. In exercise 1, the electrons are in the solid phase metal bar, for exercises 2 and 3, the electrons are in the solid Pt metal.

5.

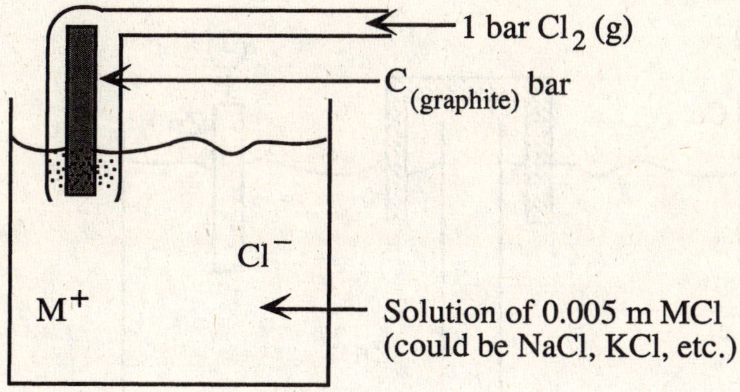

6. $Sn^{2+}_{(aq)}$, $Sn^{4+}_{(aq)} | Pt$

7.

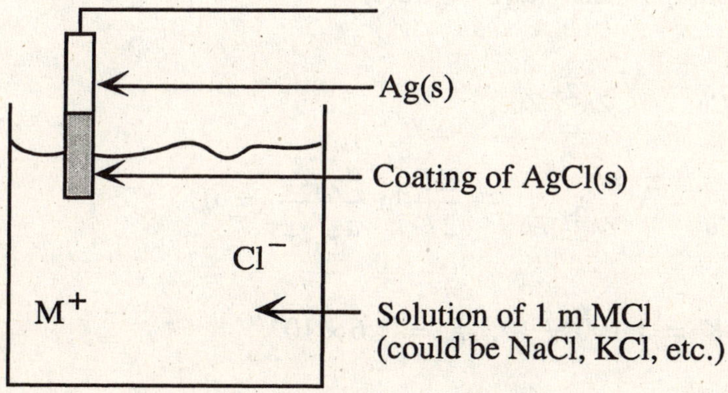

CA E5 Electrochemical Cells

Exercises

1. $$\nu_{Cu^{2+},Cu} = \nu^{\theta}_{Cu^{2+},Cu} - \frac{RT}{2F} \ln \frac{1}{a_{Cu^{2+}}}$$

$$\nu_{Zn^{2+},Zn} = \nu^{\theta}_{Zn^{2+},Zn} - \frac{RT}{2F} \ln \frac{1}{a_{Zn^{2+}}}$$

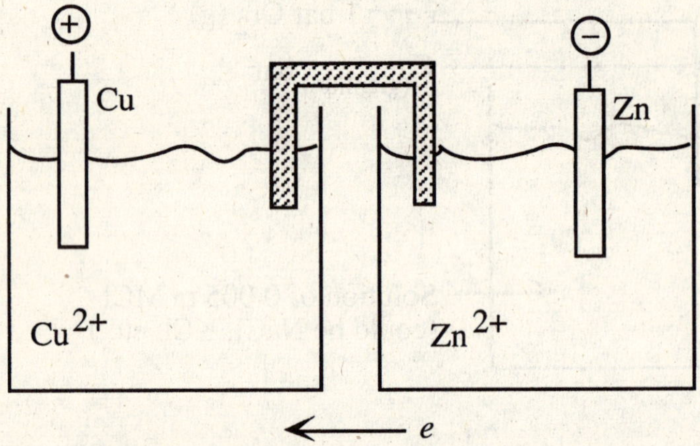

T = 298 K

$$Cu^{2+}(aq) + Zn(s) = Zn^{2+}(aq) + Cu(s)$$

$$E = \nu_{Cu^{2+},Cu} - \nu_{Zn^{2+},Zn}$$

$$E = \nu^{\theta}_{Cu^{2+},Cu} - \nu^{\theta}_{Zn^{2+},Zn} - \frac{RT}{2F} \ln \frac{a_{Zn^{2+}}}{a_{Cu^{2+}}} = 0$$

$$E^{o} = \frac{RT}{2F} \ln K = 1.100 \text{ V} \quad , \quad K = 1.6 \times 10^{37}$$

2.

	ν^{θ}(V)
$AgBr(s) + e^- = Ag(s) + Br^-(aq)$	0.0713
$Ag(s) = Ag^+(aq) + e^-$	–0.7991
$AgBr(s) = Ag^+(aq) + Br^-(aq)$	–0.728

$$\Delta G^{\theta} = -nFE^{\theta} = -RT \ln K \quad , \quad K = 4.9 \times 10^{-13}$$

3.

	V^θ(V)
$AgCl(s) + e^- = Ag(s) + Cl^-(aq)$	0.2223
$Zn^{2+}(aq) + 2e^- = Zn(s)$	–0.7628

Cell reaction:

$2\ AgCl(s) + Zn(s) = 2\ Ag(s) + Zn^{2+}(aq) + 2\ Cl^-(aq)$

$E^\theta = 0.2223 - (-0.7628) = 0.9851$

$$E_{cell} = E^\theta - \frac{RT}{2F}\ln\left(a_{Zn^{2+}}\,a^2_{Cl^-}\right)$$

$$1.1566 = 0.9851 - \frac{8.314(298)}{2(96{,}485)}\ln\left(\gamma_\pm^3\, m_{Zn^{2+}}\, m^2_{Cl^-}\right)$$

$m_{Zn^{2+}} = 0.01021, \quad m_{Cl^-} = 0.02042$

$\gamma_\pm = 0.72$

4. $$E = E^\theta_{Ag^+,Ag} - \frac{RT}{F}\ln\frac{a_{H^+}}{a_{Ag^+}}$$

neg. $H^+(aq) + e^- = \frac{1}{2}H_2(g)$

pos. $Ag+(aq) + e^- = Ag(s)$

$$E = E^\theta_{Ag^+,Ag} + \frac{RT}{F}\ln a_{Ag^+} = E^\theta_{Ag^+,Ag} + \frac{RT}{F}\ln\left(\gamma_+ m_{Ag^+}\right)$$

$$E - \frac{RT}{F}\ln m_{Ag^+} = E^\theta_{Ag^+,Ag} + \frac{RT}{F}\ln\gamma_+ = E^\theta_{Ag^+,Ag} + \frac{RT}{F}(-1.17)\,I^{1/2}$$

$I = m$

$$E - 0.0591\log m_{Ag^+} = E^\theta_{Ag^+,Ag} - 0.026\,m^{1/2}$$

E^θ is the intercept of a straight line plot of $E - 0.0591\log m_{Ag^+}$ vs $m^{1/2}$.

5.

	V^θ(V)
$2\,H^+(aq) + 2\,e^- = H_2(g)$	0.00
$AgCl(s) + e^- = Ag(s) + Cl^-(aq)$	0.2223

Spontaneous cell reaction:

$$2\,AgCl(s) + H_2(g) = 2\,H^+(aq) + 2\,Ag(s) + 2\,Cl^-(aq)$$

$$E_{cell} = E^\theta - \frac{RT}{2F}\ln a^2_{H^+}a^2_{Cl^-} = E^\theta - \frac{RT}{F}\ln a_{H^+}a_{Cl^-}$$

$$E_{cell} = E^\theta - \frac{2RT}{F}\ln \gamma_\pm - \frac{RT}{F}\ln m_{H^+}m_{Cl^-} = E^\theta - \frac{2RT}{F}\ln \gamma_\pm - \frac{2RT}{F}\ln m$$

$$m_\pm^\nu = m^2 = m_+^{\nu_+}\,m_-^{\nu_-}$$

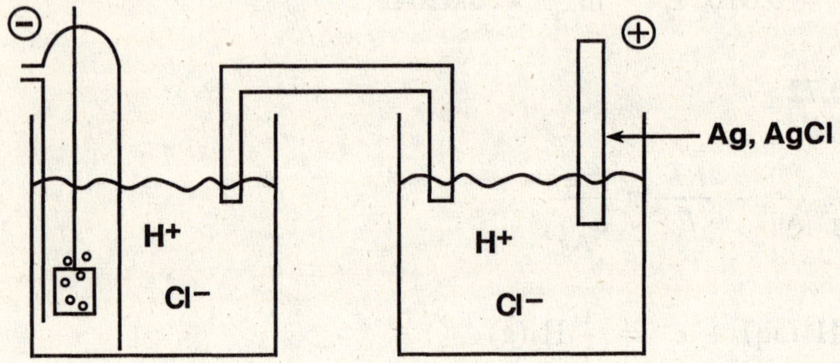

If the AgCl electrode is selected as the right-hand electrode $E^\theta > 0$ and $\Delta G^\theta < 0$. Unless concentrations are too far removed from the standard state, selection of the electrode with the most positive V^θ will lead to the spontaneous reaction by subtracting $V_R - V_L$.

6. Consult a table of standard electrode potentials. The electrode with the most positive V^θ will be the positive electrode, the electrode toward which electrons will move in the external circuit. This electrode will be the cathode.

Cathode
Cu^{2+}/Cu
H^+/H_2
Fe^{3+}/Fe^{2+}

7. $\mathrm{V}^{\theta}_{Zn^{2+},Zn} = -0.763$ $\qquad$ $\mathrm{V}^{\theta}_{H^{+},H_2} = 0.00$

$$\mathrm{V}^{\theta}_{Zn^{2+},Zn} = \frac{\mu^{\theta}_{Zn^{2+}}}{2F}$$

$$\mu^{\theta}_{Zn^{2+}} = -147 \text{ kJ mol}^{-1} \;, \qquad \mu^{\theta}_{H^{+},H_2} = 0.00$$

The zinc ion has a greater tendency to dissolve off the zinc bar and leave electrons on the bar. Thus, the zinc bar is negative relative to the hydrogen cell.

8. Yes, a potential would be developed if $m_1 \neq m_2$. This is known as a concentration cell.

The half-cell reaction for both electrode is:

$2\,H^{+}(aq) + 2\,e^{-} = H_2(g)$, for which

$$\mathrm{V} = \mathrm{V}^{\circ} - \frac{RT}{2F} \ln \frac{P_{H_2}}{\gamma_{\pm} m^2_{H^{+}}}$$

If the molalities are different, V will be different, $E = \mathrm{V}_2 - \mathrm{V}_1 \neq 0$.

CA E6 Temperature Dependence of Cell Voltage

Exercises

1. The cell reaction is:

$$O_2(g) + 2\,H_2O(l) + 2\,H_2(g) = 4\,OH^-(aq) + 4\,H^+(aq)$$

From the standard Gibbs energy and enthalpy data:

$$\Delta_r G^\theta = -154 \text{ kJ/mol}$$
$$\Delta_r H^\theta = -350 \text{ kJ/mol}$$
$$\Delta_r S^\theta = -658 \text{ J K}^{-1} \text{ mol}^{-1}$$

$$\left.\frac{\partial E^\theta}{\partial T}\right)_P = \frac{\Delta_r H^\theta}{nFT} + \frac{nFE^\theta}{nFT} = \frac{-350\times10^3}{4(96{,}485)298} + \frac{0.410}{298} = 1.70\times10^{-3} \text{ V/K}$$

2. $\Delta_r H^\theta = -nFE^\theta + nFT \left.\frac{\partial E^\theta}{\partial T}\right)_P$

$$= -2(96{,}485)(0.2676) + 2(96{,}485)(-3.19\times10^{-4}) = -7.00\times10^4 \text{ J/mol}$$

$$\Delta_r G^\theta = -nFE^\theta = -5.2\times10^4 \text{ J/mol}$$

$$\Delta_r S^\theta = nF\left.\frac{\partial E^\theta}{\partial T}\right)_P = -61.6 \text{ J K}^{-1} \text{ mol}^{-1}$$

CA K1 Introduction to Chemical Kinetics

Exercises

1. $2\,H_2(g) + O_2(g) \;=\; 2\,H_2O(g)$

$$-\frac{d(H_2)}{dt} = 1.64 \times 10^{-5}\ \frac{\text{mol}}{\text{hr}}$$

a) $$-\frac{1}{2}\frac{d(H_2)}{dt} = \frac{d(O_2)}{dt} = \text{rate of reaction}$$

$$-\frac{d(O_2)}{dt} = \frac{1}{2}(1.64 \times 10^{-5}\ \text{mol/hr}) = 0.820 \times 10^{-5}\ \text{mol/hr}$$

b) rate of reacton $= 0.820 \times 10^{-5}$ mol/hr

2. False. The reactants will be consumed at equal rates only if the stoichiometric coefficients are equal.

3. a) $3\,ClO^-(aq) \;=\; 2\,Cl^-(aq) + ClO_3^-(aq)$

b) $$\frac{d\xi}{dt} = \frac{d(CO_3^-)}{dt} = \frac{0.20\ \text{mol/L}}{100\ \text{s}} = 2.0 \times 10^{-3}\ \frac{\text{M}}{\text{s}}$$

4. $$\frac{\text{initial rate}_2}{\text{initial rate}_1} = \frac{2.70 \times 10^{-7}}{1.35 \times 10^{-7}} = 2.00 = \frac{k(0.100)^x\,(0.010)^y}{k(0.100)^x\,(0.0050)^y}$$

$$2.00 = \left(\frac{0.010}{0.0050}\right)^y = 2^y \quad \therefore\ y = 1$$

$$\frac{\text{initial rate}_3}{\text{initial rate}_2} = \frac{5.40 \times 10^{-7}}{2.70 \times 10^{-7}} = 2.00 = \frac{k(0.200)^x\,(0.010)^y}{k(0.100)^x\,(0.010)^y}$$

$2.00 = 2^x \quad \therefore\ x = 1$

reaction rate $= k(NH_4^+)\,(NO_2^-)$

5. at $t = 0$ $\quad (NO_2^-)_0 = 0.005$ M $\quad (NH_4^+)_0 = 0.100$ M

at $t = 75{,}000$ s $\quad (NO_2^-) \approx 0.0006$ M $\quad (NH_4^+) \approx 0.0956$ M ≈ 0.10 M

$$\frac{\text{rate at t} = 75{,}000}{\text{rate at t} = 0} = \frac{1.80 \times 10^{-8}}{1.35 \times 10^{-7}} = \frac{k(0.0006)^y\,(0.010)^x}{k(0.005)^y\,(0.010)^x}$$

$$0.13 = \left(\frac{0.0006}{0.005}\right)^y = (0.12)^y$$

$\therefore\ y = 1$

The order of reacton with respect to NH_4^+ cannot be determined from this graph because the conc. of NH_4^+ does not change appreciably.

6. a) For experiments 1 and 2:

$$\frac{(NO)_2}{(NO)_1} = \frac{8.07 \times 10^{-3}\ M}{5.38 \times 10^{-3}\ M} = 1.50 \qquad \frac{\text{init rate}_2}{\text{init rate}_1} = \frac{2.15 \times 10^{-5}}{9.55 \times 10^{-6}} = 2.25$$

↑ inc in NO ↑ inc in rate

$(1.50)^2 = 2.25$

$\therefore$ rate of reaction $= k(NO)^2\,(O_2)^x$

For experiments 4 and 5:

$$\frac{(O_2)_5}{(O_2)_4} = \frac{9.69 \times 10^{-3}\ M}{6.99 \times 10^{-3}\ M} = 1.39$$

$$\frac{\text{init rate}_5}{\text{init rate}_4} = \frac{3.88 \times 10^{-5}}{2.80 \times 10^{-5}} = 1.39$$

$\therefore$ rate of reaction $= k(NO)^2\,(O_2)$

b) init rate$_1$ $= k(5.38 \times 10^{-3}\ M)^2\,(5.38 \times 10^{-3}\ M)$

$= 9.55 \times 10^{-6}$ M/s

$$k = \frac{9.55 \times 10^{-6}\ M/s}{1.56 \times 10^{-7}\ M^3} = \frac{6.13 \times 10^1}{M^2 s}$$

(c) $$\text{init rate} = \frac{6.13 \times 10^1}{M^2 s}\,(1.50 \times 10^{-2}\ M)^2\,(2.00 \times 10^{-3}\ M)$$

init rate $= 2.76 \times 10^{-5}$ M/s

CA K2 Integrated Rate Laws

Exercises

1. rate of reaction $= -\dfrac{d(B)}{dt} = k(B)^2$

$$\frac{d(B)}{(B)^2} = -k\,dt \quad \int_{(B)_o}^{(B)} \frac{d(B)}{(B)^2} = -\int_0^t k\,dt$$

$$-\frac{1}{(B)} + \frac{1}{(B)_o} = -kt$$

$$\frac{1}{(B)} = \frac{1}{(B)_o} + kt$$

2. (a) For a first-order reaction, rate = k (A). The units are $\frac{M}{s} = k\,M$. k must have units s^{-1}.

(b) For a second-order reaction, rate = k $(A)^2$. The units are $\frac{M}{s} = k\,M^2$. k must have units $M^{-1}\ s^{-1}$.

(c) For rate = k $(A)^{1/3}$ (B). The units are $\frac{M}{s} = k\,M^{4/3}$. k must have units $M^{-1/3}$ s^{-1}.

3. Determine the initial concentration of A, $(A)_o$. Begin the reaction and measure the concentration of A as a function of time. Note the time when $(A) = \frac{1}{2}(A)_o$; this is $t_{1/2}$. Repeat the experiment with a new initial concentration of A, $(A)_o'$; measure the new half-life, $t'_{1/2}$. If the reaction is first-order, $t_{1/2}$ equals $t'_{1/2}$. If the reaction is second-order, the value of k and k' are equal, where

$$k = \frac{1}{t_{1/2}\,(A)_o} \quad \text{and} \quad k' = \frac{1}{t'_{1/2}\,(A)_o'}$$

4. I — No, (N_2O_4) is increasing!
   II — No, (N_2O_4) does not decrease linearly for 1st order
   III — No, (N_2O_4) = constant!
   IV — Yes, exponential decay of (N_2O_4)
   V — No, (N_2O_4) is decreasing, but not exponentially

CA K3 The Method of Isolation for Determining Rate Laws

Exercises

1. We don't know $(H_2O_2)_o$. But, we do know that all of the H_2O_2 produces 57.90 mL of O_2 at $t = \infty$. Therefore, we will assume

$$(H_2O_2)_t \propto [V_{O_2} \text{ at } t = \infty] - [V_{O_2} \text{ at time } t] = 57.90 \text{ mL} - [V_{O_2} \text{ at time } t]$$

If the reaction is 1st order, plot of $\ln (V_\infty - V_t)$ vs. t will be a straight line. The plot is a straight line, see below.

Slope = -2.76×10^{-2} m^{-1}

$$k_{app} = \frac{2.76 \times 10^{-2}}{\text{min}}$$

$H_2O_2 + I^-$ /Data

time (min)	volume O_2 (mL)	$V_\infty - V_t$ (mL)	$\ln (V_\infty - V_t)$
0.00	0.00	57.90	4.05871738
5.00	7.50	50.40	3.91999118
10.00	14.00	43.90	3.78191432
25.00	28.80	29.10	3.37073817
45.00	41.20	16.70	2.81540872
65.00	48.30	9.60	2.2617631
∞	57.90	0.00	

Hydrogen Peroxide Decomposition in the Presence of Iodide Ion

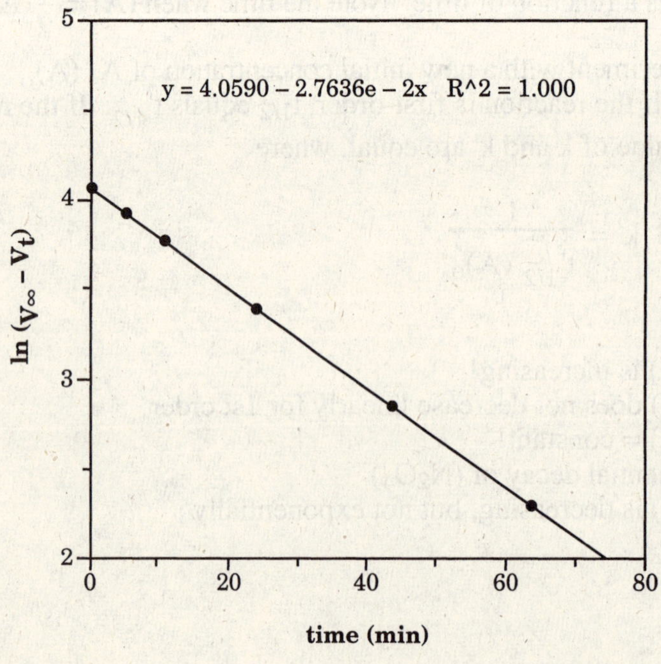

2. If the reaction is 1st order,

$$(HCrO_4^-) = (HCrO_4^-)_o\, e^{-kt} = C_o\, e^{-kt}$$

at $t = 0$ $(HCrO_4^-) = C_o$ and $[Cr^{3+}] = 0$

at $t = t$ $(HCrO_4^-) = C_o e^{-kt}$ and $(Cr^{3+}) = C_o - (HCrO_4^-)$

$$[Cr^{3+}] = C_o - C_o e^{-kt} = C_o\,(1 - e^{-kt})$$

CA K4 Reaction Mechanism

Exercises

1. a) $\text{rate}_3 = k_3(A)(B)$

b) $-\dfrac{d(A)}{dt} = \text{rate}_3$

c) $\dfrac{d(C)}{dt} = \text{rate}_3$

2. a) $\text{rate}_2 = k_2(A)^2$

b) $-\dfrac{1}{2}\dfrac{d(A)}{dt} = \text{rate}_2$ $\qquad -\dfrac{d(A)}{dt} = 2 \times \text{rate}_2$

c) $\dfrac{d(C)}{dt} = \text{rate}_2$

3. rate of production of CH_4 = $\dfrac{d(CH_4)}{dt} = k_2(CH_3)(C_2H_6)$

CH_3 is produced in step 1 $\quad \dfrac{1}{2}\dfrac{d(CH_3)}{dt} = k_1(C_2H_6)$

CH_3 is consumed in step 2 $\quad -\dfrac{d(CH_3)}{dt} = k_2(CH_3)(C_2H_6)$

Steady-State for CH_3

$$2k_1(C_2H_6) = k_2(CH_3)(C_2H_6) \qquad \text{eqn I}$$

$$(CH_3) = \frac{2k_1}{k_2}$$

$$\frac{d(CH_4)}{dt} = k_2(CH_3)(C_2H_6) = 2k_1(C_2H_6)$$

---

rate of production of H_2 = $\dfrac{d(H_2)}{dt} = k_4(H)(C_2H_6)$

Steady-State for H

$$\frac{d(H)}{dt} = k_3\,(C_2H_5) \qquad -\frac{d(H)}{dt} = k_4\,(H)(C_2H_6)$$

$$k_3(C_2H_5) = k_4(H)(C_2H_6) \qquad \text{eqn II}$$

$$(H) = \frac{k_3\,(C_2H_5)}{k_4(C_2H_6)} \qquad \text{eqn III}$$

Try C_2H_5

$$\frac{d(C_2H_5)}{dt} = k_2(CH_3)(C_2H_6) + k_4(H)(C_2H_6)$$

$$-\frac{d(C_2H_5)}{dt} = k_3(C_2H_5) + 2k_5\,(C_2H_5)^2 + 2k_6\,(C_2H_5)^2$$

$$k_2(CH_3)(C_2H_6) + k_4(H)(C_2H_6) = k_3(C_2H_5) + 2k_5(C_2H_5)^2 + 2k_6(C_2H_5)^2$$

Use equations I and II

$$2k_1(C_2H_6) + k_4(H)(C_2H_6) = k_4(H)(C_2H_6) + 2k_5(C_2H_5)^2 + 2k_6(C_2H_5)^2$$

$$2k_1(C_2H_6) = 2(k_5 + k_6)\,(C_2H_5)^2$$

$$(C_2H_5)^2 = \frac{k_1}{k_5 + k_6}\,(C_2H_6)$$

$$(C_2H_5) = \left(\frac{k_1}{k_5 + k_6}\right)^{1/2} (C_2H_6)^{1/2}$$

Use equation III

$$(H) = \left(\frac{k_3}{k_4\,(C_2H_6)}\right)\left(\frac{k_1}{k_5 + k_6}\right)^{1/2} (C_2H_6)^{1/2}$$

$$(H) = \frac{k_3}{k_4}\left(\frac{k_1}{k_5 + k_6}\right)^{1/2} (C_2H_6)^{-1/2}$$

$$\frac{d(H_2)}{dt} = \mathrm{k}_4\,\frac{k_3}{k_4}\left(\frac{k_1}{k_5 + k_6}\right)^{1/2} (C_2H_6)^{-1/2}\,(C_2H_6)$$

$$= k_3 \left(\frac{k_1}{k_5 + k_6} \right)^{1/2} (C_2H_6)^{1/2}$$

4. a) Mechanism A

$$\text{rate} = \frac{1}{2} \frac{d(NO_2)}{dt} = k_{A2}\,(N_2O_2)(O_2)$$

$$\frac{d(N_2O_2)}{dt} = k_{A1}\,(NO)^2$$

$$-\frac{d(N_2O_2)}{dt} = k_{-A1}(N_2O_2) + k_{A2}\,(N_2O_2)(O_2)$$

$$k_{A1}\,(NO)^2 = (N_2O_2)\,\{k_{A1} + k_{A2}\,(O_2)\}$$

$$\text{rate} = \frac{k_{A2}\,k_{A1}\,(NO)^2(O_2)}{k_{-A1} + k_{A2}\,(O_2)}$$

acceptable when (if) $k_{-A1} \gg k_{A2}\,(O_2)$

Mechanism B

$$\text{rate} = \frac{1}{2} \frac{d(NO_2)}{dt} = k_{B2}\,(NO_3)(NO)$$

$$\frac{d(NO_3)}{dt} = k_{B1}\,(NO)\,(O_2)$$

$$-\frac{d(NO_3)}{dt} = k_{-B1}(NO_3) + k_{B2}\,(NO_3)(NO)$$

$$k_{B1}\,(NO)\,(O_2) = (NO_3)\,\{k_{-B1} + k_{B2}\,(NO)\}$$

$$\text{rate} = \frac{k_{B2}\,k_{B1}\,(NO)^2(O_2)}{k_{-B1} + k_{B2}\,(NO)}$$

acceptable when (if) $k_{-B1} \gg k_{B2}\,(NO)$

(b) Run an experiment where (O_2) is very high. If mechanism A is correct, reaction will be zero order with respect to O_2.

Run an experiment where (O_2) is very high. If mechanism B is correct, reaction will be first order with respect to NO (and first order with respect to O_2).

5. CH_4 is produced in step 2 only:

$$\frac{d(CH_4)}{dt} = k_2\,(CH_3)(CH_3CHO)$$

C_2H_6 is produced in step 4 only:

$$\frac{d(C_2H_6)}{dt} = k_4\,(CH_3)^2$$

Steady-State to CH_3

$$-\frac{d(CH_3)}{dt} = k_2\,(CH_3CHO)(CH_3) + 2k_4\,(CH_3)^2$$

$$\frac{d(CH_3)}{dt} = k_1\,(CH_3CHO) + k_3\,(CH_2CHO)$$

$$k_2\,(CH_3CHO)\,(CH_3) + 2k_4\,(CH_3)^2 = k_1\,(CH_3CHO) + k_3\,(CH_2CHO)$$

Steady-State to CH_2CHO

$$k_2\,(CH_3)\,(CH_3CHO) = k_3\,(CH_2CHO)$$

$$\therefore\ k_2\,(CH_3CHO)\,(CH_3) + 2k_4\,(CH_3)^2 = k_1\,(CH_3CHO) + k_2\,(CH_3)\,(CH_3CHO)$$

$$2k_4(CH_3)^2 = k_1(CH_3CHO)$$

$$(CH_3) = \left(\frac{k_1}{2k_4}\right)^{1/2} (CH_3CHO)^{1/2}$$

$$\frac{d(CH_4)}{dt} = k_2 \left(\frac{k_1}{2k_4}\right)^{1/2} (CH_3CHO)^{3/2}$$

$$\frac{d(C_2H_6)}{dt} = \frac{k_4 k_1}{2k_4}\,(CH_3CHO)$$

6\. a) $$\text{rate} = \frac{d(P)}{dt} = k_2'\,(B)(A)$$

$$\frac{d(B)}{dt} = k_1(A)^2$$

$$-\frac{d(B)}{dt} = k_{-1}(B) + k_2'\,(B)(A)$$

$$k_1(A)^2 = k_{-1}(B) + k_2'\,(B)(A) = (B)\,\{k_{-1} + k_2'(A)\}$$

$$(B) = \frac{k_1\,(A)^2}{k_{-1} + k_2'(A)}$$

$$\therefore\ \text{rate} = \frac{k_2' k_1\,(A)^3}{k_{-1} + k_2'(A)}$$

b) If the reaction is second order with respect to A and the alternative mechanism is correct, then $k_2'(A)$ must be much larger than k_{-1}.

A good experiment would be to run the reaction with a very small conc. of A, so that

$$k_2'(A) \ll k_{-1}$$

Then, the reaction should be third order with respect to A.

CA K5 Activation Energy (I)

Exercises

1.

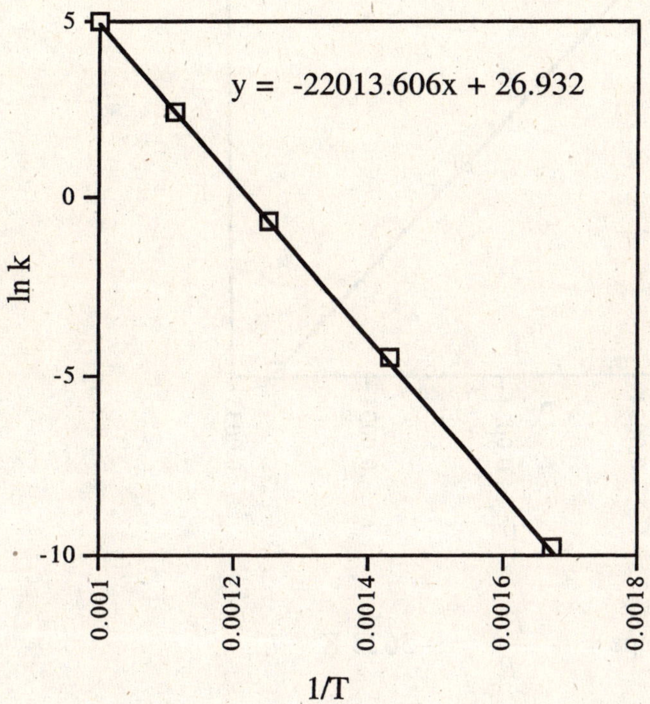

$$\text{slope} = -\frac{E_a}{R} = -22014\ \text{K}$$

$E_a = +22014\ \text{K} \times R = +22014\ \text{K} \times 8.314 \frac{\text{J}}{\text{mol K}} = 1.83 \times 10^5 \frac{\text{J}}{\text{mol}} = 183$ kJ/mol

intercet = $\ln A = 26.9$ $\qquad A = 4.8 \times 10^{11}\ \text{s}^{-1}$

2.

T (°C)	Chirps per minute	T (K)	$1/T$	ln chirp rate
25.0	178	298.2	3.354×10^{-3}	5.182
20.3	126	293.5	3.407×10^{-3}	4.836
17.3	100	290.5	3.442×10^{-3}	4.605

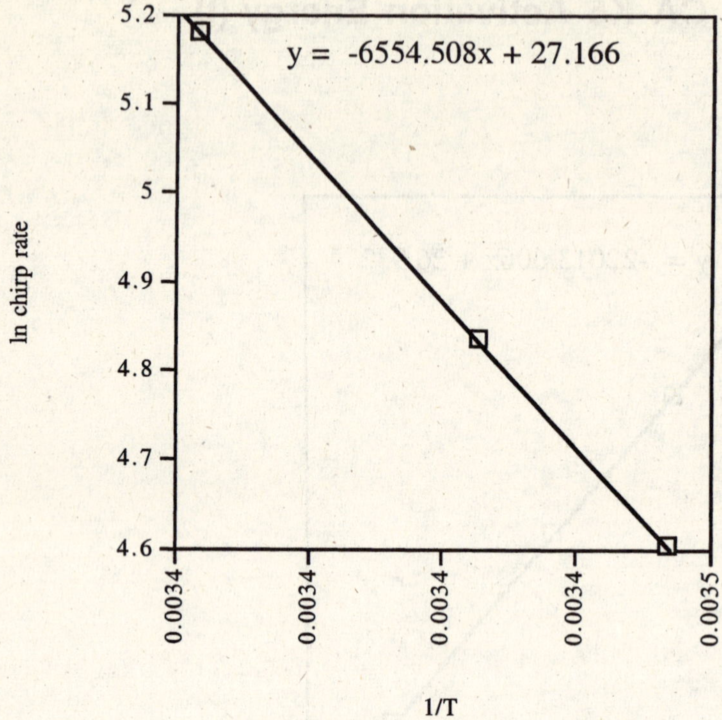

$$\text{slope} = -\frac{E_a}{R} = -5623 \text{ K}$$

$E_a = +6555 \text{ K} \times R = +6555 \text{ K} \times 8.314 \frac{\text{J}}{\text{mol K}} = 5.45 \times 10^4 \frac{\text{J}}{\text{mol}} = 54.5$ kJ/mol

intercept = ln A = 27.17 $\quad A = 6.31 \times 10^{11}\ \text{s}^{-1}$

$k_{14°\text{C}} = A\, e^{-E_a/RT} = 6.31 \times 10^{11}\ \text{s}^{-1}\ e^{-5.45 \times 10^4/8.314\,(287)} = 76$ chirps/minute

At 14°C, the number of chirps in 15 s = 19

19 + 40 = 59°F (not bad for a cricket)
°F = 1.8 × 14 + 32 = 57

3. $$\frac{k_{35}}{k_{25}} = \frac{Ae^{-E_a/R(308)}}{Ae^{-E_a/R(298)}} = 3 = \frac{e^{-E_a/R(308)}}{e^{-E_a/R(298)}}$$

$$3\,e^{-E_a/R(298)} = e^{-E_a/R(308)}$$

$$\ln 3 - \frac{E_a}{298\,R} = -\frac{E_a}{308\,R} \qquad \ln 3 = \frac{E_a}{R}\left(\frac{1}{298} - \frac{1}{308}\right)$$

$$E_a = \frac{\ln 3 \times R}{0.109 \times 10^{-3}} = \frac{1.099 \times 8.314}{0.109 \times 10^{-3}} \approx 8 \times 10^4 \text{ J/mol} \approx 80 \text{ kJ/mol}$$

CA K5A Activation Energy (II)

Exercises

1. $$k = \left(\frac{A_1 A_3 A_4}{A_5}\right)^{1/2} (C_2H_6) \quad A_i = e^{-E_{a_i}/RT}$$

$$E_a(\text{overall}) = \frac{1}{2}(351 + 167 + 29 - 0) = 274 \text{ kJ/mol}$$

2. $$K_{298} = \frac{k_1}{k_{-1}} = \frac{0.10}{2.0 \times 10^{-3}} = 50$$

$$\ln\frac{k_{313}}{k_{298}} = -\frac{E_{a_f}}{R}\left(\frac{1}{313} - \frac{1}{298}\right) = \ln\frac{0.25}{0.10}$$

$E_{a_f} = 474$ kJ/mol
Similarly $E_{a_r} = 358$ kJ/mol

$\Delta_r H = E_{a_f} - E_{a_r} = 116$ kJ/mol

3. The overall activation energy is negative, which is very unusual. This could occur if the concentration of an intermediate species decreases as the temperature increases. A lower concentration of an intermediate species would result in a lower rate. In this case (N_2O_2) decreases with increasing temperature.

CA K6 Enzyme Kinetics

Exercises

1. $$(E{\bullet}S) = \frac{k_1\,(S)(E)}{k_{-1} + k_2} = \frac{k_1\,(S)\,\{(E)_{tot} - (E \bullet S)\}}{k_{-1} + k_2} = \frac{k_1\,(S)\,(E)_{tot} - k_1\,(E \bullet S)}{k_{-1} + k_2}$$

$$[k_{-1} + k_2]\,(E{\bullet}S) = k_1\,(S)\,(E)_{tot} - k_1\,(S)\,(E{\bullet}S)$$

$$[k_{-1} + k_2 + k_1\,(S)]\,(E{\bullet}S) = k_1\,(S)\,(E)_{tot}$$

$$(E{\bullet}S) = \frac{k_1\,(S)\,(E)_{tot}}{k_{-1} + k_2 + k_1\,(S)}$$

2.

Conc of CO_2 (mmol/L)	Initial rate (mmol/Ls)	1/Conc of CO_2 (L/mmol)	1/Initial rate (Ls/mmol)
0.76	0.0062	1.316	161.29
1.51	0.0116	0.662	86.21
3.78	0.0204	0.265	49.02
7.57	0.0287	0.132	34.84
15.1	0.0368	0.066	27.17

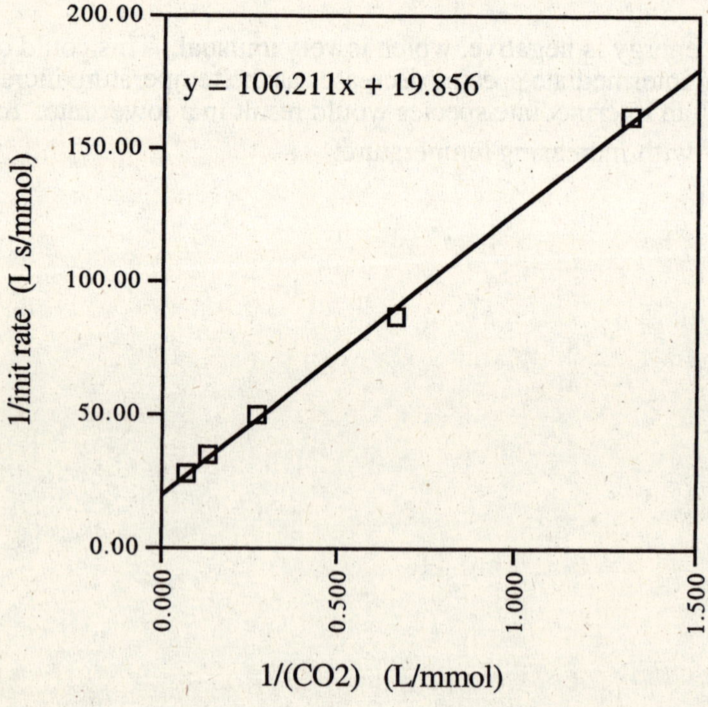

$$\text{intercept} = \frac{1}{k_2\,(E)_{tot}} = 19.9\,\frac{\text{L s}}{\text{mmol}} \qquad k_2\,(E)_{tot} = 0.0503\,\frac{\text{mmol}}{\text{Ls}}$$

$$\text{slope} = \frac{K_M}{k_2\,(E)_{tot}} = 106\ \text{s}$$

$$K_M = 106\ \text{s} \times\ 0.0503\ \frac{\text{mmol}}{\text{Ls}} = 5.33\ \frac{\text{mmol}}{\text{L}}$$

(Recall that K_M is the conc of substrate which yields a rate of reaction equal to $\frac{1}{2}$ rate$_{max}$.

3. If (S) = K_M

$$\text{rate} = \frac{k_2\,(S)(E)_{tot}}{2\,(S)} = \frac{1}{2}\,k_2\,(E)_{tot} = \frac{1}{2}\,\text{rate}_{max}$$

4. $k = A\,e^{-Ea/RT}$ Assume A is unchanged by the catalyst.

$$\frac{k_{cat}}{k_{uncat}} = \frac{e^{-E_{acat}/RT}}{e^{-E_{auncat}/RT}} = \frac{e^{-8\times10^3/RT}}{e^{-75.3\times10^3/RT}} = e^{67.3\times10^3/(8.314\times300)} = 5\times10^{11}$$

5. $\Delta_rH = E_{a_f} - E_{a_r}$, Δ_rH is unaffected by a catalyst.

$$E_{a_f} - E_{a_r} = 120 - 185 = -65\ \text{kJ/mol}$$

$$-65 = 90 - E_{a_r}, \qquad E_{a_r} = 155\ \text{kJ/mol}$$

6. The substrate could undergo direct conversion to product or a catalyst could be used to produce an intermediate species. The catalyst might encapsulate the substrate and hold it in an advantageous position for reaction to occur.

7.

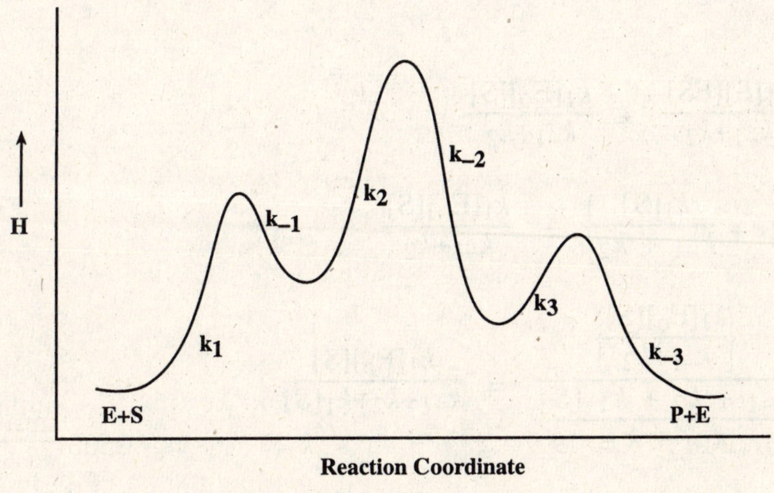

$$E + S \underset{k_{-1}}{\overset{k_1}{\rightleftharpoons}} ES \underset{k_{-2}}{\overset{k_2}{\rightleftharpoons}} EP \underset{k_{-3}}{\overset{k_3}{\rightleftharpoons}} P + E$$

ES = enzyme-substrate complex

EP = enzyme product

ES high reactive

$k_2 >> k_{-2}$, $k_3 >> k_{-3}$

EP decays rapidly

$$V = \frac{d[P]}{dt} = -\frac{d[S]}{dt} = k_2[ES]$$

k_2 is assumed to be rate limiting.

$$\frac{d[ES]}{dT} = 0 = k_1[E][S] - k_{-1}[ES] - k_2[ES]$$

Now $[E] = [E_o] - [ES] - [EP]$

EP dissociates rapidly and is neglected.

$$[E] \cong [E_o] - [ES]$$

$$\frac{d[ES]}{dT} = k_1[S]\ \{[E_o - [ES]\}\ - k_{-1}\ [ES] - k_2\ [ES] = 0$$

$$[ES] = \frac{k_1[E][S]}{k_{-1}+k_2} = \frac{k_1[S]}{k_{-1}+k_2}\ \{[E_o - [ES]\}$$

$$= \frac{k_1[E_o][S]}{k_{-1}+k_2} - \frac{k_1[S][ES]}{k_{-1}+k_2}$$

$$[ES] + \frac{k_1[S][ES]}{k_{-1}+k_2} = \frac{k_1[E_o][S]}{k_{-1}+k_2}$$

$$[ES] \times \left\{1 + \frac{k_1[S]}{k_{-1} + k_2}\right\} = \frac{k_1[E_o][S]}{k_{-1}+k_2}$$

$$[ES] = \frac{\left\{\frac{k_1[E_o][S]}{k_{-1}+k_2}\right\}}{\frac{k_{-1} + k_2 + k_1\ [S]}{k_{-1} + k_2}} = \frac{k_1[E_o][S]}{k_{-1}+k_2+k_1[S]}$$

$$k_1[S][E_o] - k_1[S][ES] - k_{-1}[ES] - k_2[ES] = 0$$

$$[ES] = \frac{k_1[E_o][S]}{k_1[S] + k_{-1} + k_2}$$

$$V = \frac{k_2 k_1[E_o][S]}{k_1[S] + k_{-1} + k_2} = \frac{k_2[E_o][S]}{[S] + \frac{k_{-1}+k_2}{k_1}}$$

V_{max} will occur when all the enzyme is in the form of ES complex.

$\therefore$

$$V_{max} = k_2[E_o]$$

define

$$K_m = \frac{k_{-1} + k_2}{k_1}$$

then

$$V = \frac{V_{max}[S]}{K_m + [S]}$$

CA K6A Collision Theory of Gas Phase Reactions

Exercises

1. $Z_1 = \sqrt{2}\ \bar{v}\ \pi\ d^2\ N/V$

$$\bar{v} = \left(\frac{8RT}{\pi M}\right)^{1/2} = 4.2 \times 10^2\ \text{ms}^{-1}$$

$$\frac{P}{kT} = \frac{N}{V} = \frac{(1\ \text{bar})(10^5\ \text{Pa/bar})}{(1.381\times10^{-23}\ \text{J/K})(273\ \text{K})} = 2.7 \times 10^{25}\ \text{molecules/m}^3$$

$$Z_1 = \sqrt{2}\ \pi\ (2.7 \times 10^{25})(3.6\times 10^{-10})^2\ (4.2 \times 10^2) = 6.5 \times 10^9\ \text{collisions/m}^3\text{s}$$

$$Z_{11} = \frac{1}{2} Z_1 \frac{N}{V} = 8.8 \times 10^{34}\ \text{collisions/m}^3\text{s}$$

2. The mean free path, λ, is given by the average distance traversed in unit time, $\bar{v}$ divided by the number of collisions experienced,

$$\lambda = \frac{\bar{v}}{Z_1} = \frac{\bar{v}}{\sqrt{2}\ \pi \frac{N}{V} d^2 \bar{v}} = \frac{1}{\sqrt{2}\ \pi \frac{N}{V} d^2}$$

$$\frac{N}{V} = \frac{P}{kT}, \qquad \lambda = \frac{kT}{\sqrt{2}\ \pi P d^2}$$

Increase in T increases λ, increase in P decreases λ.

For oxygen at O°C and 1 bar,

$$\lambda = \frac{(1.38\times10^{-23})(273)}{\sqrt{2}\ \pi (3.6\times10^{-10})^2 (1.00\times10^5)} = 6.6 \times 10^{-8}\ \text{m}$$

3. $Z_1 = \sqrt{2}\ \bar{v}\ \pi\ d^2\ N/V$

If V doubles, Z_1 is halved.

$$Z_{11} = \frac{1}{2}(\sqrt{2}\ \bar{v}\ \pi\ d^2\ N/V)\ N/V$$

If V doubles, Z_{11} is lowered by a fourth.

CA K6B Analysis of a Gas Phase Chain Reaction

Exercises

1. $$\frac{d(\mathrm{HBr})}{dt} = k_2(\mathrm{Br})(\mathrm{H_2}) + k_3(\mathrm{H})(\mathrm{Br_2}) - k_4(\mathrm{HBr})(\mathrm{H})$$

$$= k_2\left(\frac{k_1}{k_5}\right)^{1/2}(\mathrm{Br_2})^{1/2}(\mathrm{H_2}) + \frac{k_3(\mathrm{Br_2})k_2\left(\frac{k_1}{k_5}\right)^{1/2}(\mathrm{Br_2})^{1/2}(\mathrm{H_2})}{k_3(\mathrm{Br_2}) + k_4(\mathrm{HBr})}$$

$$- \frac{k_4(\mathrm{HBr})k_2\left(\frac{k_1}{k_5}\right)^{1/2}(\mathrm{Br_2})^{1/2}(\mathrm{H_2})}{k_3(\mathrm{Br_2}) + k_4(\mathrm{HBr})}$$

$$= k_2\left(\frac{k_1}{k_5}\right)^{1/2}(\mathrm{Br_2})^{1/2}(\mathrm{H_2})\left\{1 + \frac{k_3(\mathrm{Br_2}) - k_4(\mathrm{HBr})}{k_3(\mathrm{Br_2}) + k_4(\mathrm{HBr})}\right\}$$

$$= k_2\left(\frac{k_1}{k_5}\right)^{1/2}(\mathrm{Br_2})^{1/2}(\mathrm{H_2})\left\{\frac{k_3(\mathrm{Br_2}) + k_4(\mathrm{HBr}) + k_3(\mathrm{Br_2}) - k_4(\mathrm{HBr})}{k_3(\mathrm{Br_2}) + k_4(\mathrm{HBr})}\right\}$$

$$= k_2\left(\frac{k_1}{k_5}\right)^{1/2}(\mathrm{Br_2})^{1/2}(\mathrm{H_2})\left\{\frac{2k_3(\mathrm{Br_2})}{k_3(\mathrm{Br_2}) + k_4(\mathrm{HBr})}\right\}$$

$$= \left\{\frac{2k_2\left(\frac{k_1}{k_5}\right)^{1/2}k_3(\mathrm{Br_2})^{1/2}(\mathrm{H_2})(\mathrm{Br_2})}{k_3(\mathrm{Br_2}) + k_4(\mathrm{HBr})}\right\}$$

$$= \left\{\frac{2k_2\left(\frac{k_1}{k_5}\right)^{1/2}(\mathrm{Br_2})^{1/2}(\mathrm{H_2})}{1 + \frac{k_4(\mathrm{HBr})}{k_3(\mathrm{Br_2})}}\right\}$$

2. a) $\Delta_r H^\circ$ = 436 kJ/mol for the reaction given (from values in Table 1). The reverse reaction, $2\,\mathrm{H} + \mathrm{M} \rightarrow \mathrm{H_2} + \mathrm{M}$, has $E_a \approx 0$. Therefore, $E_{a_r} \approx 436$ kJ/mol.

b) The rate constant for the reaction $\mathrm{Br_2} + \mathrm{M} \xrightarrow{k_1} 2\,\mathrm{Br} + \mathrm{M}$ is much larger than for the reaction in part (a).

$$\frac{k_6}{k_1} = \frac{A_6}{A_1}\frac{e^{-E_6/RT}}{e^{-E_1/RT}} \approx e^{-(436-193)10^3/RT} \approx 10^{-43} \qquad \text{at 300 K.}$$

c) Using the steady state approximations:

$$\frac{(\mathrm{H})}{(\mathrm{Br})} \approx \frac{k_2\,(\mathrm{H_2})}{k_3\,(\mathrm{Br_2}) + k_4\,(\mathrm{HBr})} \approx \frac{k_2}{k_3} \text{ at the beginning of the reaction}$$

$$\frac{(\mathrm{H})}{(\mathrm{Br})} \approx e^{-(70\times10^3)/RT} = 10^{-13}$$

The concentration of Br is more than a billion times higher than that of H.

3. The reverse of step 3 is not competitive with the forward reaction. E_{a_3} is only 4 kJ/mol whereas E_{a_8} is at least 177 kJ/mol. For reaction 3 $\Delta_r H°$ is –173 kJ/mol,:

$$\Delta_r H = E_{a_f} - E_{a_r} = 4 - E_{a_r}, \qquad E_{a_r} = 177 \text{ kJ/mol}$$

4. $E_a = E_{a_2} + \frac{1}{2}E_{a_1} - \frac{1}{2}E_{a_5} = 74 + 193/2 - 0 = 171$ kJ/mol

5. Steps 2 and 3 are propagation steps. These steps produce intermediates that give rise to further intermediates. Step 4 is inhibition. This step slows the reaction by removing the product.

CA K7 Transition State Theory

Exercises

1. a) $$\frac{d\ln K_c^{\ddagger}}{dT} = \frac{d\left(-\Delta G^{\circ\ddagger}/RT\right)}{dT} = \frac{\Delta H^{\circ\ddagger}}{RT^2}$$

$$k_r = K_c^{\ddagger}\frac{kT}{h}$$, take ln of the equation

$$\ln k_r = \ln K_c^{\ddagger} + \ln T + \ln\frac{k}{h}$$

b) From Arrhenius,

$$\ln k_r = \ln A - \frac{E_a}{RT}$$

$$\frac{d(\ln k_r)}{dT} = \frac{E_a}{RT^2}$$

c) From transition state theory (a) above

$$\frac{d(\ln k_r)}{dT} = \frac{1}{T} + \frac{d\ln K_c^{\ddagger}}{dT}$$

d) Combining (b) and (c)

$$\frac{E_a}{RT^2} = \frac{1}{T} + \frac{d\ln K_c^{\ddagger}}{dT}$$

e) $$\frac{E_a}{RT^2} - \frac{1}{T} = \frac{d\ln K_c^{\ddagger}}{dT} = \frac{\Delta H^{\circ\ddagger}}{RT^2}$$

$$\Delta H^{\circ\ddagger} = E_a - RT$$

2. $$\ln\frac{k_{r473}}{k_{r373}} = -\frac{E_a}{R}\left(\frac{1}{473} - \frac{1}{373}\right)$$, $E_a = 1.6 \times 10^2$ kJ/mol

$$\Delta H^{\circ\ddagger} = E_a - RT = 1.6 \times 10^2 \text{ kJ/mol}$$

$$k_{373} = \frac{kT}{h}e^{-\Delta G^{\circ\ddagger}/RT} = \frac{1.38\times10^{-23}\,(373)}{6.62\times10^{-34}}e^{-\Delta G^{\circ\ddagger}/RT}$$

$$\Delta G^{\circ\ddagger}_{373} = 1.4 \times 10^2 \text{ kJ/mol}$$

$$\Delta G^{\circ\ddagger} = \Delta H^{\circ\ddagger} - T\,\Delta S^{\circ\ddagger}\ ; \qquad \Delta S^{\circ\ddagger} = 54 \text{ J/(K mol)}$$

3. $\Delta H^{\circ\ddagger} = E_a - RT = 160 \times 10^3 - 8.314(300) = 158$ kJ/mol

$$k_{r300} = 4.0 \times 10^{13}\ e^{-\Delta G^{\circ\ddagger}/R(300)} = 5.6 \times 10^{-15}\ s^{-1}$$

$$k_{r300} = \frac{kT}{h}\ e^{-\Delta G^{\circ\ddagger}/R(300)} = 5.6 \times 10^{-15}\ s^{-1}$$

$\Delta G^{\circ\ddagger} = 156$ kJ/mol , $\Delta G^{\circ\ddagger} = \Delta H^{\circ\ddagger} - T\,\Delta S^{\circ\ddagger}$; $\Delta S^{\circ\ddagger} = 7$ J/(K mol)

or, $A = 4.0 \times 10^{13} = \dfrac{kT}{h}\ e\ e^{\Delta S^{\circ\ddagger}/R}$

4. a) $$\left(\frac{\partial \ln k_r}{\partial P}\right)_T = -\frac{\Delta V^{\circ\ddagger}}{RT}$$

If $\Delta V^{\circ\ddagger} = 0$, there is no pressure effect.
If $\Delta V^{\circ\ddagger} > 0$, an increase in pressure will decrease k_r.
If $\Delta V^{\circ\ddagger} < 0$, an increase in pressure will increase k_r.

b) If $\Delta V^{\circ\ddagger} = 0$, the volume of the reactants is the same as those of the transition state. $\Delta V^{\circ\ddagger} > 0$ means the transition state volume is greater than that of the reactants.
If $\Delta V^{\circ\ddagger} < 0$, the transition state volume is less than that of the reactants.

c) If the rate increases, the rate constant increases. This means $\Delta V^{\circ\ddagger} < 0$, the transition state occupies a smaller volume than the reactants.

5. A rate increase with increasing temperature means the volume of the transition state is less than that of the reactants. A possible transition state could be $(C_2H_5I{\bullet}C_2H_5O)^-$.

6. a) $\Delta H^{\circ\ddagger} = E_a - RT = 221 \times 10^3 - 8.314(513) = 216$ kJ/mol

$$k_{r573} = \frac{kT}{h}\ e^{-\Delta G^{\circ\ddagger}/R(573)} = \frac{1.38\times10^{-23}\,(573)}{6.62\times10^{-34}}\ e^{-\Delta G^{\circ\ddagger}/R(573)}$$

$$= 1.19 \times 10^{13}\ e^{-\Delta G^{\circ\ddagger}/4.76\times10^3}$$

$$k_{r573} = 1.25 \times 10^{14}\ e^{-221{,}000/RT} = 8.9 \times 10^{-7}\ s^{-1}$$

$\Delta G^{\circ\ddagger} = 208$ kJ/mole

$\Delta G^{\circ\ddagger} = 208 \times 10^3 = 216 \times 10^3 - (573)\,(\Delta S^{\circ\ddagger})$, $\Delta S^{\circ\ddagger} = 13$ J/(K mol)

b) $\Delta S^{\circ\ddagger}$ is very small so the transition state is much like the reactant with slightly loosend bonds.

c) $\Delta V^{\circ\ddagger}$ should be only slightly positive. There will be little effect of pressure on the rate constant.

7. a) $$\ln \frac{24.95\times10^5}{1.72\times10^5} = -\frac{E_a}{R}\left(\frac{1}{318} - \frac{1}{298}\right)$$

$$E_a = 105 \text{ kJ/mol}$$

b) $$\Delta H^{\circ\ddagger} = E_a - RT = 106\times10^3 - 8.314(298) = 103 \text{ kJ/mol}$$

$$k_r = \frac{kT}{h} e^{-\Delta G^{\circ\ddagger}/RT} = 1.72\times10^{-5}$$

$$\Delta G^{\circ\ddagger} = 100 \text{ kJ/mol}$$

$$\Delta G^{\circ\ddagger} = \Delta H^{\circ\ddagger} - T\,\Delta S^{\circ\ddagger} = 100\times10^3 = 100\times10^3 - 298(\Delta S^{\circ\ddagger})$$

$$\Delta S^{\circ\ddagger} = 10 \text{ J/(K mol)}$$

c) $\Delta S^{\circ\ddagger}$ is very small so the transition state is much like the reactant with slightly loosend bonds.

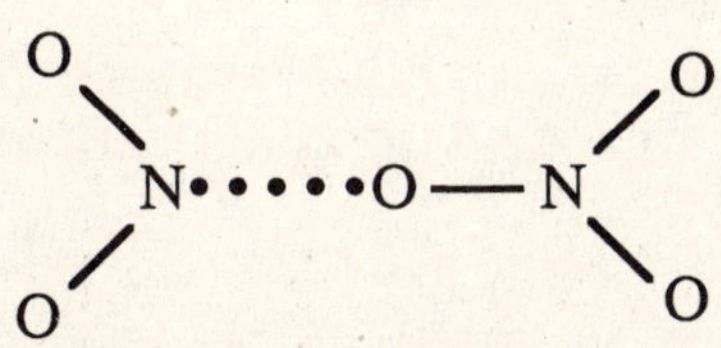

d) $\Delta V^{\circ\ddagger}$ should be only slightly positive. There will be little effect of pressure on the rate constant.

CA M1 Mathematics for Thermodynamics

Exercises

1. $\dfrac{df(x)}{dx} = f'(x) \qquad \dfrac{d^2f(x)}{dx^2} = f''(x)$

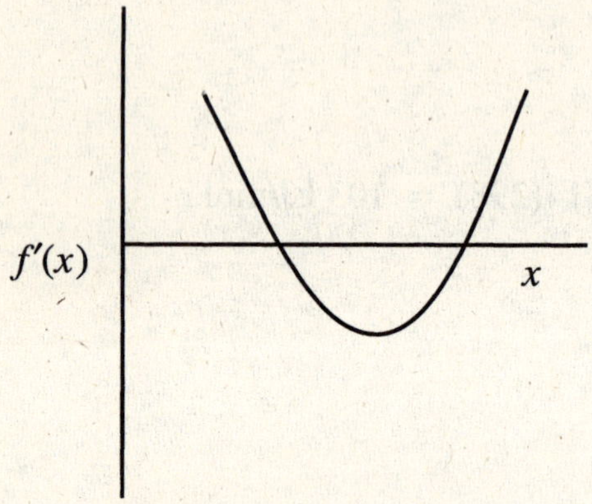

2.

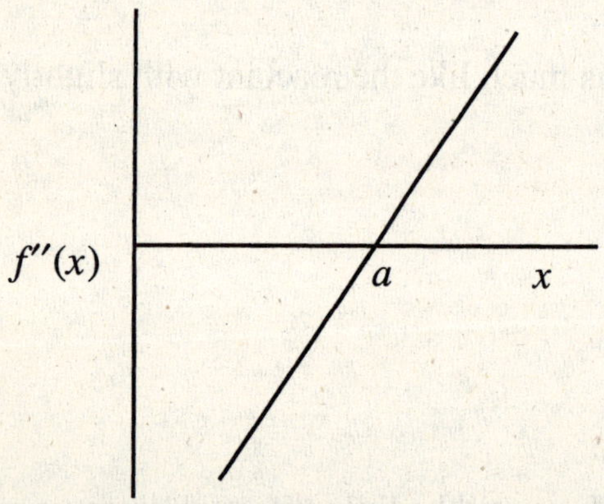

3. $f'(x) = 0$ at the maximum and minimum of $f(x)$.

4. $f'(x)$ changes from positive to negative at a maximum point of $f(x)$.

5. $f'(x)$ changes from negative to positive at a minimum point of $f(x)$.

6. If $f''(x)$ is positive the function $f(x)$ has a minimum.

7. If $f''(x)$ is negative the function $f(x)$ has a maximum.

8. If $f''(x) = 0$ and changes sign as x increases, then $F(x)$ has a point of inflection at $x = a$.

9. $y = x$, $dy = dx$, $\frac{dy}{dx} = 1$

$y = 1/x$, $dy = -1/x^2 dx$, $\frac{dy}{dx} = \frac{-1}{x^2}$

$f(x) = \frac{Z(x)}{Y(x)}$, $\frac{df(x)}{dx} = \frac{Y(x)Z'(x) - Z(x)Y'(x)}{Y(x)^2}$

$f(x) = Z(x)\, Y(x)$, $\frac{df(x)}{dx} = Z(x)\, Y'(x) + Y(x)Z'(x)$

10. $\int_{x_1}^{x_2} d\ln x = \ln x_2 - \ln x_1$, $\int_{P_1}^{P_2} \frac{dP}{P^2} = -\left(\frac{1}{P_2} - \frac{1}{P_1}\right)$

$\int_{V_1}^{V_2} dV = V_2 - V_1$, $\int_{T_1}^{T_2} R dT = R(T_2 - T_1)$

11. $\int_{V_1}^{V_2} \frac{dV}{V} = \ln V_2 - \ln V_1 = \ln V_2/V_1$

12. $\int_{V_1}^{V_2} P dV = \int_{V_1}^{V_2} \frac{RT}{V} dV$

The integral can be evaluated only if T is a constant, or if T is known as a function of volume.

13.

$$\left.\frac{\partial V}{\partial T}\right)_P = \left.\frac{\partial\left(\frac{RT}{P}\right)}{\partial T}\right)_P = \frac{R}{P} = \frac{0.08206}{\text{atm}} \frac{\text{L-atm}}{\text{K}}$$

$n=1$

$= 0.08206$ L/K

14. The volume is constant; the pressure will increase as the temperature increases. This can be expressed by the derivative $\left.\frac{\partial P}{\partial T}\right)_V$.

$$\left.\frac{\partial P}{\partial T}\right)_V = \left.\frac{\partial\left(\frac{RT}{V}\right)}{\partial T}\right)_V = \frac{R}{V} = \frac{0.08206}{1\text{L}}\ \frac{\text{L-atm}}{\text{K}}$$

n=1

$$= 0.08206\ \frac{\text{atm}}{\text{K}}$$

15.

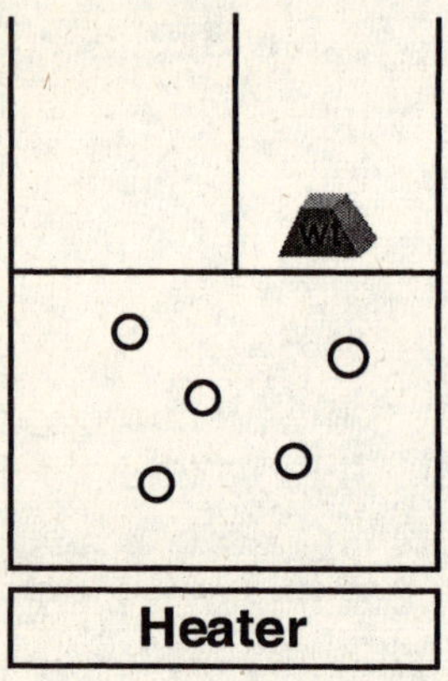

P = constant

At constant pressure the volume is changed by adding or extracting heat causing the temperature to change.

16. $$\left.\frac{\partial T}{\partial V}\right)_{n,P} = \left.\frac{\partial\left(\frac{PV}{R}\right)}{\partial V}\right)_{n,P} = \frac{P}{R} = \frac{\text{atm K}}{\text{L atm}} = \frac{\text{K}}{\text{L}}$$

n=1

17. $$\left.\frac{\partial P}{\partial T}\right)_V = \frac{R}{V} + \frac{b}{V^2}$$

Let $P = P(\text{T},V)$

$$dP = \left.\frac{\partial P}{\partial T}\right)_V dT + \left.\frac{\partial P}{\partial V}\right)_T dV$$

If P = constant

$$0 = \left.\frac{\partial P}{\partial T}\right)_V \mathrm{d}T + \left.\frac{\partial P}{\partial V}\right)_T \mathrm{d}V$$

$$\frac{\mathrm{d}V}{\mathrm{d}T} = \left.\frac{\partial V}{\partial T}\right)_P = -\frac{\partial P/\partial T)_V}{\partial P/\partial V)_T} = -\frac{\dfrac{R}{V} + \dfrac{b}{V^2}}{\dfrac{-RT}{V^2}\ \dfrac{-2a}{V^3}\ \dfrac{-2bT}{V^3}}$$

P = constant

18. $\mathrm{d}V = \left.\frac{\partial V}{\partial T}\right)_P \mathrm{d}T + \left.\frac{\partial V}{\partial P}\right)_T \mathrm{d}T$

$PV = RT$

$$\mathrm{d}V = \frac{R}{P}\,\mathrm{d}T - \frac{RT}{P^2}\,\mathrm{d}P$$

$$\left.\frac{\partial(R/P)}{\partial P}\right)_T = -\frac{R}{P^2}\ ;\quad \left.\frac{\partial\left(\frac{-RT}{P^2}\right)}{\partial T}\right)_P = \frac{-R}{P^2}$$

dV is exact.

19. $\mathrm{d}w = -P\mathrm{d}V$, using dV from Exercise 18,

$$\mathrm{d}w = -R\mathrm{d}T + \frac{RT}{P}\mathrm{d}P + \left.\frac{\partial\left(\frac{RT}{P}\right)}{\partial T}\right)_P = \frac{R}{P},\ \left.\frac{\partial R}{\partial P}\right)_T = 0$$

dw is not exact.

20. Along Path 1 the gas is heated from T_1 to T_2 at constant P_1 and then expanded from P_1 to P_2 at constant T_2.

Along Path 2 the gas is expanded from P_1 to P_2 at constant T_1 and then heated from T_1 to T_2 at constant P_2.

$$\int_{T_1,P_1}^{T_2,P_2} \mathrm{d}V = \int_{T_1,P_1}^{T_2,P_2} \left(\frac{R}{P}\,\mathrm{d}T - \frac{RT}{P^2}\,\mathrm{d}P\right)$$

Path 1

$$= \frac{R}{P_1}\int_{T_1}^{T_2} \mathrm{d}T - RT_2 \int_{P_1}^{P_2} \frac{\mathrm{d}P}{P^2}$$

$$= \frac{R}{P_1}(T_2 - T_1) - RT_2\left(\frac{1}{P_1} - \frac{1}{P_2}\right)$$

$$= \frac{RT_2}{P_1} - \frac{RT_1}{P_1} - \frac{RT_2}{P_1} + \frac{RT_2}{P_2}$$

$$= V_2 - V_1$$

$$\int_{T_1,P_1}^{T_2,P_2} dV = \int_{T_1,P_1}^{T_2,P_2} \left(\frac{R}{P}\,dT - \frac{R\,T}{P^2}\,dP\right)$$

Path 2

$$= -\int_{P_1}^{P_2} RT_1 \frac{dP}{P^2} + \int_{T_1}^{T_2} \frac{R}{P_2}\,dT$$

$$= RT_1\left(\frac{1}{P_2} - \frac{1}{P_1}\right) + \frac{R}{P_2}(T_2 - T_1)$$

$$= \frac{RT_1}{P_2} - \frac{RT_1}{P_1} + \frac{RT_2}{P_2} - \frac{RT_1}{P_2}$$

$$= V_2 - V_1$$

$$\int_{T_1,P_1}^{T_2,P_2} dw = \int_{T_1,P_1}^{T_2,P_2} \left(\frac{RT}{P}\,dP - R\,dT\right)$$

Path 1

$$= -\int_{T_1}^{T_2} R\,dT + \int_{P_1}^{P_2} \frac{RT_2}{P}\,dP$$

$$= -R(T_2 - T_1) + RT_2 \ln\frac{P_2}{P_1}$$

$$\int_{T_1,P_1}^{T_2,P_2} \mathrm{d}w = \int_{T_1,P_1}^{T_2,P_2} \left(\frac{RT}{P}\,\mathrm{d}P - R\,\mathrm{d}T\right)$$

Path 2

$$= \int_{P_1}^{P_2} RT_1\,\frac{\mathrm{d}P}{P} - \int_{T_1}^{T_2} R\mathrm{d}T$$

$$= RT_1 \ln\frac{P_2}{P_1} - R(T_2 - T_1)$$

21. The change in volume depends only on the initial volume, V_1, and the final volume, V_2. ΔV is the same along Path 1 or Path 2 as it is along any path. Volume is an exact differential.

The work done depends on the path. The work along Path 1 is not the same as the work along Path 2. Work is not an exact differential.